Investment Projects in Agriculture

Principles and Case Studies

McDonald P. Benjamin

Longman

Longman Group Limited
Longman House,
Burnt Mill,
Harlow, Essex, U.K.

First published 1981

British Library Cataloguing in Publication Data

Benjamin, McDonald P
Investment projects in agriculture.
1. Underdeveloped areas-Agriculture-Economic aspects-1945-
1. Title
338.1'09172'4 HD1417

ISBN 0 582 64306 8 ✓ 1/9/82 see back page

Printed and bound in Great Britain by
William Clowes (Beccles) Limited, Beccles and London

Contents

List of Figures

List of Tables

This book is dedicated to:

my wife, Hannelore
my son, McDonald (Jr), and
my daughter, Sandra.

Acknowledgement

The writer is grateful to Mr Jan P. Huyser, formerly Director of FAO's Investment Centre for the encouragement he gave me in undertaking the writing of this book. Much basic material from the Investment Centre especially in two case studies in Part IV of the text was used in the process, and this is gratefully acknowledged.

No meaningful attempt to discuss the topic of agricultural investment projects can be made without reference to the World Bank's experience in this field, since this Bank has the broadest and deepest experience available in the field of lending to agriculture. I am, therefore, particularly grateful to the World Bank for permission to use its material. This relates especially to the Nigeria Case Study and the Mexico Rural Development Project discussed in the text.

I have benefited considerably from the comments of both Mr J. Price Gittinger of the Economic Development Institute of the World Bank, and Mr John Clark, of the Investment Centre, who kindly read the first draft. Special thanks are due to Dr Andrew MacMillan, also of the Investment Centre, who laboured with me through all the various drafts of the text, and contributed substantially to this book through intensive discussions of the more substantive issues.

It should be emphasised, however, that the views expressed in this book are solely the responsibility of the writer. They do not necessarily reflect the views of the Food and Agriculture Organisation of the United Nations (FAO), of which the writer is an official, nor do they represent the views of the World Bank, with whom the writer was associated through the FAO/World Bank Cooperative Programme. Lastly, they do not represent the views of the International Fund for Agricultural Development, IFAD, where the writer now works.

The typing of the manuscript was done by Mrs Anna Lisa Cibien-Colizzi and Mrs Aurora Robino, and their contribution is gratefully acknowledged.

Above all, my largest thanks go to my wife, Hannelore Benjamin, who not only assisted in typing the earliest drafts but also sacrificed so many weekends and holiday periods while I wrote this book. Her patience and determination to see this project through have, in fact, made this book possible. In this process, my children also shared in the sacrifice.

Foreword

It is with great pleasure that I have accepted to write this foreword to Dr M. P. Benjamin's book, since he has been associated with IFAD since its inception.

The International Fund for Agricultural Development (IFAD) has now had three years' experience in the execution of its mandate, since its launching in December 1977. The uniqueness of its structure is equally matched by the uniqueness of its mandate, endorsed fully by its 135 member Governments.

In the context of North–South relations, IFAD is the most visible symbol of the type of cooperation towards which developing countries wish to move so as to have greater involvement in global decisions affecting their welfare. IFAD's constitution permits the beneficiaries to sit as equals with the donors – the OECD and OPEC countries – in its governing bodies.

Adam Smith's 1776 statement: 'No society can surely be flourishing and happy, of which by far the greater part of the numbers are poor and miserable', provides the logical framework of IFAD's predominant concentration on its poorest member countries and the poverty groups in these countries so as to involve them more fully in food production and the development process. In fact, the problems of income distribution and poverty redressal are further accentuated as *per capita* food production decreases in the developing countries and upward pressures on food prices continue in the absence of a dependable system of food security.

What lessons have we learnt as the newest international entrant in the field of agricultural development and lending? The first is that our special mandate to increase food production and farm incomes with primary attention to small and landless farmers, requires a recognition of the fact that the problem of hunger and malnutrition in the developing countries is rooted in the broader problem of rural poverty.

The problems are not intractable but require that Governments, in their formulation of agricultural projects, introduce special measures to reflect these concerns. Examples from our own experience are quite promising. Thus, IFAD has sought to create an asset base for the small farmers and landless by, for example, getting governments to agree that any new or improved land would be made available to the small farmers and landless; more equitable crop sharing arrangements between landlords and tenants; establishing explicit lending targets for small farmers in rural credit schemes, and the inclusion of special project components to enable the rural poor to purchase assets for engaging in non-farm production activities.

On the institutional side we have learnt that a major reason why the poorest populations fail to receive their fair share in the economic development process, is the absence of efficient rural institutions through which the benefits of projects can be channelled to them. IFAD projects pay special attention to this aspect.

Despite our achievements, we face severe constraints at two levels that affect the pace and direction of our operations. The first relates to the acute shortage of skills in project preparation in many of our member countries. Closely related to this is the problem of the required expertise to implement projects approved for financing.

The problem, therefore, is one of manpower development, and the training implications of this enormous task. It is in this context that Dr M. P. Benjamin's book, written in his personal capacity, has a special contribution to make. His message is clear: planners in the developing countries must themselves become more fully involved in charting development strategies and investment projects appropriate to their specific country requirements. The highly legible form in which this book is written, and its wide-ranging practical examples, should ensure a very wide usage. The book, therefore, provides a very useful tool for training of nationals in the developing world in the difficult field of preparing agricultural projects for financing, with or without external assistance.

Abdelmuhsin M. Al-Sudeary
President, International Fund for Agricultural Development

Rome, March 1981.

Introduction

This book has the primary objective of assisting students and practitioners of agricultural development in developing countries to enhance their own capacity for designing, preparing and implementing agricultural projects. The multilateral and bilateral donors, in their aid to agriculture, require that projects meet certain minimum standards to qualify for financing.

Students of agriculture and agricultural economics in the developing world concentrate on single disciplines, such as agronomy or livestock, or general agriculture. There is virtually no course work in the analysis of agricultural projects. Students from developing countries who attend universities and agricultural colleges in the developed countries may never be exposed to ideas on agricultural project analysis. This is largely because the 'project approach' to agricultural development, later discussed, was not the main path followed by the western industrialised countries in their own process of agricultural development.

There are two main reasons why agricultural project analysis has not received the required attention among students in the developing world. First, there is virtually complete unawareness of the opportunities that exist for teaching institutions to contribute to their countries' project approach to development. Knowledge of the importance and role of projects is limited to a small circle of government officials in key ministries, who negotiate loans with outside donors, and those directly involved in project implementation. Second, there is a marked absence of course material in this field of project analysis as the recognised existing literature concentrates primarily on the economic analysis of projects, which is only one of the many aspects of project formulation.

With the sharply expanding role of the project approach to agricultural development in the developing countries, the increasing need for qualified manpower in project preparation and implementation is being filled, in the large bulk of the developing countries, by technical assistance, which is now becoming the privileged domain of expatriate consulting firms and individual consultants. The sharply increasing costs in recent years of employing such outside consultants is a response, not to inflation, but to supply and demand. In fact, in some of the poorest African countries, the technical assistance component for project implementation may reach as high as 25 per cent of total project costs. In fact, 10–20 per cent is not an uncommon figure. Unfortunately, this is generally not a free gift to the recipient country since it has to be paid back to the donor.

Yet, for many of the more sophisticated governments in the developing world, the question of technical assistance for project preparation and implementation is becoming a thorny issue. There are questions of national pride, and resistance to this enclave of foreign experts who enjoy salaries and privileges that bear no relationship at all to the inferior working conditions of their counterparts in the recipient country. The response of some governments to this challenge is the establishment of project preparation units or cells, whose task it is to prepare projects both for financing with external support and for exclusive domestic financing. A case in point is India where the target is to bring 1·5 million ha under irrigation each year for the next five years; India could not achieve this without its own project-preparation capacity. Experienced local consulting firms also exist, which are fully aware of the requirements of the financing institutions.

The multilateral agencies recognise this problem of deficient project-preparation capacity. Examples are: the World Bank through its Economic Development Institute regularly conducts training courses in project preparation for mainly government officials from developing countries. The Food and Agriculture Organisation of the United Nations (FAO) periodically holds such courses. The Institute of Social Studies of the Hague also conducts such courses. The United Nations Development Programme (UNDP) is paying increasing attention to giving assistance to countries to build project-preparation teams. This approach is quite feasible on a regional basis also, as in Central America.

The missing links in this chain of events are the agricultural and economics teaching institutions in the developing world. They have the lessons of experience all around them in the form of on-going financed projects. As a first step, they can monitor and evaluate the performance and impact of these projects.

Staff and students of these faculties could, by arrangements with the government agencies concerned, directly participate in the work of the visiting missions and/or consultants for project identification and preparation. This approach would have the added benefit that local teaching institutions could build up a series of case studies as a basis for teaching project-analysis work.

Investment requirements in agriculture

A key indicator of the magnitude of the task of agricultural development facing the developing countries is the projected level of investment requirements. The proposals, considered by the World Food Conference at its Rome meeting in 1974, called for a total investment of US$120 thousand million (billion), for the period 1975–85 at 1972 prices, or roughly an annual average of US$11 billion for the period. The projected cereal gap for the developing countries would reach 85 million tonnes involving a drain of foreign exchange of US$14 to 16 billion annually. Historic growth rates of 2·6 per cent for food and agricultural production in these countries would have to rise to new levels of 3·5–4·0 per cent on an annual basis.

The Conference argued that the US$1·5 billion of the current (1972) annual external aid flow for food production would have to rise to US$5·3 billion per year in 1972 prices in the period 1975–80. In 1975 prices, the requirement would be US$8·3 billion.

Other agencies and institutions have made estimates of requirements on the basis of varying assumptions. The International

Food Policy Research Institute study, limited to 36 developing countries with severe food deficits, estimated that annual investments of US$5·7 billion in 1975 prices are required to overcome their food deficit. Such agencies as the World Bank, FAO and the World Food Council (a product of the World Food Conference) have also examined the aspect of investment requirements. They all conclude that aid-flows, especially on concessional terms, have to be considerably stepped up if targets for increases in agricultural and food production are to be met by the developing countries themselves.

The main areas where investment should be directed include irrigation and water development, first and foremost, especially in Asia. The development of rainfed agriculture including infrastructure such as roads and electrification, is important. Other fields include research, extension and training; the provision of basic inputs such as seeds, fertilisers, pesticides and mechanisation, and the reduction of post-harvest losses through storage and drying.

Levels and trends in aid-flows to agriculture

In response to the calls from the World Food Conference to the international community, two notable developments have occurred: first, the very substantial increase in aid-flows to agriculture primarily from the established multilateral agencies, and second, the establishment of the International Fund for Agricultural Development (IFAD), both discussed below.

The World Food Council which has, as a major objective, the task of inducing larger aid-flows to agriculture, has recently examined these trends, based on data from the Development Assistance Committee (DAC) of OECD and other sources. The DAC classifies aid-flows to agriculture into two main classes: namely, the narrow definition with investments geared more directly to food and agricultural production, while the broad definition includes such investments as agro-industries, rural infrastructure and fertiliser plants.

On this basis, aid-flows to agriculture (narrow definition) increased from US$2·53 billion in 1973 to US$4·30 billion in 1977, all expressed in 1975 prices. The respective figures employing the broad definition were US$3·16 billion and US$6·03 billion for the same period, also in 1975 prices. Thus substantial increases have occurred in real terms. The three major groups of donors include the multilateral, the DAC bilateral/EEC, and the OPEC bilateral. The most striking feature is the increasingly dominant role of the multilateral agencies in the agricultural lending field. The share of the multilateral lending agencies based on 1975 prices increased from about 39 per cent in 1973 to 66 per cent of lending in 1977 under the narrow definition, and from about 41 per cent to 64 per cent in the same period under the broad definition.

The credit for shaping this new trend goes primarily to the World Bank. The other multilateral lending agencies include the regional development banks for Asia, Latin America and Africa, the European Development Fund and the UNDP. In 1977, the World Bank's share of total multilateral assistance was over 70 per cent and about 40 per cent of all assistance from all donors in 1977. The 1977 commitments for food production of about US$4·3 billion in 1975 prices was still only about one half of the estimated annual resource requirements of US$8·3 billion earlier discussed.

According to the World Bank's annual report for 1978, the Bank's lending for agriculture and rural development projects reached the impressive level of US$3·27 billion during fiscal 1978 (1st July, 1977–30th June, 1978), in contrast to US$0·95 billion in fiscal 1974. Aggregate lending for agriculture and rural development totalled US$10·0 billion for a total of 363 projects over the period 1974–8. Of the Bank's total lending for all sectors, the agriculture and rural development sector increased its share from 22 per cent in fiscal 1974 to 39 per cent in fiscal 1978.

An important element of this growth in lending is the expanding share of lending for rural development projects which rose from US$0·45 billion in 1974 to US$1·73 billion in 1978, or about 53 per cent of the Bank's lending for agriculture and rural development. In this context, rural development projects include those projects in which more than half of the project's direct benefits are expected to accrue to the rural poor. This policy shift in levels and direction of the World Bank's lending derive largely from Mr McNamara's address to the Bank's Board of Governors at its 1973 Nairobi meeting, in which the Bank's concern for the eradication of rural poverty was stressed. Simultaneously, production of food assumed major emphasis in the World Bank's rural development strategy.

The second development of tremendous potential significance for the developing countries was the establishment of the International Fund for Agricultural Development at the beginning of 1978. This new specialised agency, under the United Nations system, is the first concrete expression of the New International Economic Order at work. A key distinguishing feature is the equal distribution of voting rights among the three categories, namely Category I, the established OECD donors; Category II, the new OPEC donors and Category III, the recipient countries. The initial capitalisation of US$1·0 billion was contributed 57 per cent by Category I and 43 per cent by Category II, with the recipients having made voluntary contributions equivalent to about US$20 million.

The IFAD's mandate and its published lending policies and criteria require it to concentrate about two-thirds of its resources on the poorest food-deficit countries, with increased food production, increasing employment in rural areas, and improved nutrition as the major objectives of its lending. In its first year of operations, 1978, the Fund financed ten projects totalling about US$120 million, and more than tripled its volume of lending during 1979, bringing its commitments to about US$500 million for the two-year period.

Programme and project lending

External aid to agriculture takes place almost exclusively in the form of *project lending* that is lending for specific and discrete projects. There are many advantages to this approach to lending; there is the discipline involved in the various steps of the project cycle, such as identifying, preparing and appraising the project. Strong control can be exerted during implementation by the process of supervision. Projects are useful tools to assist and influence a country's investment strategy. They can direct resources towards specific target groups and such objectives as the correction of regional income disparities in a country. They are disadvantageous in that they are time-consuming in preparation and in implementation, and tie up a considerable amount of the country's limited specialised skills in agricultural development. Generally, disbursements, except for agricultural credit projects, tend to lag behind the targets, and so the rate of absorption of a country for external financial resources for projects is much lower than would be the case with, say, *programme lending*.

Programme loans and/or credits so far granted by the international financing institutions have been sharply limited both in terms of share of their overall operations and in terms of their objectives. So far, they have taken the form mainly of providing balance-of-payments support for imports of critically needed goods, such as machinery and spares, and are limited to the poorest countries. Situations that justify programme loans by the World Bank, for example, include reconstruction or rehabilitation; a sudden fall in export earnings of an important export item; a sharp deterioration of the terms of trade of a country and structural imbalances, that may require an effective stabilisation programme. Yet, there are encouraging signs of a broadened interpretation of programme lending for agriculture. The World Bank recently approved a credit through its affiliate the International Development Association (IDA) to Bangladesh for fertiliser imports, and other lending institutions are examining similar proposals. This represents a highly progressive forward step, which can have a quick and immediate impact on agricultural production and lessen the dependence of highly vulnerable countries on food aid. The difficulties of programme lending are that their results may be difficult to monitor, and the beneficiaries of such lending may not be disadvantaged target groups.

Lastly, there are the opportunities for increased sectoral lending. For example, for irrigation development in a country or its political sub-divisions, lending for the sector is possible, provided that certain prerequisites are met. Thus, the overall development strategy for the sector over a stated time-frame must be agreed; design standards for construction and criteria for operation and maintenance of the project works must be agreed, and also the methodology for appraisal of the sub-projects involved in the programme for the sector. There is considerable merit in further development of this approach, e.g. in livestock, fisheries and rural development, that would pave the way for larger flows of aid resources to the developing countries.

Constraints to increased agricultural lending

Among the more significant constraints are: the absence of a proven technical package around which sound project formulation can proceed. This is especially the case for rainfed or non-irrigated agriculture. In Indonesia, government efforts to strongly increase transmigration from the inner to the outer islands face serious technical land-use problems. Applied research based on farming systems is sadly lacking.

There is a marked shortage of resource surveys or feasibility studies as a basis for financing. With more donors coming into the picture, e.g. several institutions based in the Middle East, there is a scramble for projects. Feasibility studies require time and resources, and many countries are not financially capable of undertaking them with their own resources. The UNDP has been a major contributor of assistance for such studies. In fact, the more obvious or readily accessible areas would have already been tapped.

The lack of project preparation capacity in the bulk of the countries has already been referred to. Sound project preparation always entails a multidisciplinary approach, as all project components must be mutually supportive of each other. With the new style projects which have other objectives besides aggregate production increases, project preparation, involving such diverse components as social services, irrigation, roads, mechanisms for beneficiary participation and for coordination, is an increasingly difficult task. These projects require more time, more qualified staff and better conceptualisation. The shortage of manpower becomes a more serious constraint when project implementation starts. Shifts of staff away from their regular assignments, provide the basis for manning new projects. There are usually no plans for training the required personnel in advance for staffing the new projects.

The absence of supportive government policies may prove a constraint. If the country does not have a rural development strategy – that is, it does not actively promote larger resource flows to the rural poor – then formulation and implementation of rural development projects in this context may prove difficult, since there must be a political commitment on the part of government, if the project's objectives are to be achieved. Marketing and pricing policies that discriminate against the producer, in favour of the urban consumer, are also a case in point.

Lastly, there is the very important consideration of local cost financing. As the countries move away from the highly capital-intensive type of projects, involving hardware, into projects that are more labour intensive and involve more software, then the project's foreign exchange component decreases and the local-cost component increases. Thus a shift in focus from civil works and structures to a people-oriented approach has immediate implications for local-cost financing. On the other hand, many of the lending institutions, especially the regional ones, have highly restrictive limitations on the extent to which they can finance the local-cost components of projects. The World Bank has shown great flexibility in this regard. IFAD has a tremendous advantage in that its lending policies confer on it considerable leeway to respond to the needs of the country and the project.

On the other hand, the severe budgetary problems that face the poorest and more vulnerable countries, mean that at periods of acute financial stress, the local matching resources planned in the budget for the project fail to materialise. It is not uncommon, therefore, that governments constantly aim at getting the lending institutions to finance larger shares of total project cost, which results in an increased contribution of foreign exchange for financing local costs.

The main contents of the text

The book is divided into four parts. Part I deals with the project cycle, project types and rural development. The project cycle takes the reader through all the various steps such as definition of project concept, identification, preparation, appraisal and loan negotiations, including the putting together of the financing package. The various steps are fully substantiated with actual examples. Reasons for undertaking projects are discussed, and a grouping of the main project classes is proposed, based on the 'objective of change' that the various projects seek to induce. Since rural development projects are now becoming so important, the experience of a large-scale project in Mexico financed by the World Bank is reviewed. The country's approach to rural development, organisation and derivation of criteria for selecting micro-regions and developing investment packages are discussed.

Part II attempts to develop a set of principles that could be applied to project formulation work in most developing countries. Projects should have clearly defined objectives, even where potential conflicts in objectives could arise. For this reason, it is necessary to build flexibility into projects. However, these objectives must be reconciled somehow with the country's power structure, and they cannot be realised if the incentives offered by the project are inadequate to overcome the elements of risk that it introduces. Projects always confront planners with a host of issues that must be clearly identified and resolved. These issues cover both economic and non-economic ones, such as organisation and management. Some guidelines for deriving the appropriate scale of the project are proposed. All projects should have built-in control or monitoring systems to serve the project's management so as to ensure realisation of the stated objectives. More attention needs to be paid in project formulation to the replicability of the project so that similar projects can later be formulated to benefit new groups. The planner must ensure adequate operating capital, and establish the true costs of the project to the country.

Part III covers the main considerations in project design and analysis. The project's technical basis is discussed and examples are given of how a technical package is derived for the agricultural-development aspects of an irrigation project in Burma, and for an integrated rural development project in Mali. The assumptions in deriving the technical package are clearly shown. The institutional features cover the advantages and disadvantages of various governmental and quasi-public institutions as they relate to projects. Boards, project authorities and corporate entities are discussed. Organisation and management (O and M) aspects are critically important for the project's success. The O and M arrangements are discussed for a rural-development project in the Philippines.

Financial and commercial aspects are discussed in terms of the objectives of the financial analysis, and the derivation of farm models. A detailed example is given of the steps involved in preparing a dairy farm model for a typical 6-ha farm in Ruritania, and how the financial rate of return is derived. The derivation of the rate of return for the milk-processing enterprise is also shown. The conventional economic analysis is discussed against a background of the newly emerging analytical tools. The derivation of the economic rate of return for an irrigation project in India is demonstrated. Land-use options in the Nepalese Terai are discussed to demonstrate the limitations of the economic rate of return as a decision-making tool.

Lastly, Part IV of the text takes the reader through three case studies. The African region is represented by a Nigerian case study, which demonstrates how a project can introduce technology into a stagnant agriculture and benefit masses of farmers. The approach lends itself to application in many African countries. The Asian region is represented by the Indonesian case study on seeds. The approach here is to demonstrate the various considerations in the project-formulation process and the rationale behind the inclusion of various project components, and the design of the organisation and management arrangements for the project. The Latin American and Caribbean region is represented by the Jamaican agricultural credit case study. The two main considerations of a credit programme are discussed, namely, the lending institution and the lending programme.

The writer has attempted in this book to strike a judicious balance between broad principles that should guide project planners, such as setting the scale of a project, and the need to maximise the use of actual examples covering a wide range of projects in an equally wide range of countries. The objectives of this book shall have been well served if it creates greater understanding among planners and students in the developing world, of the role they can play in the process of agricultural development through the instrument of agricultural investment projects.

The writer's own experience in one developing country, when he worked as agricultural adviser to the government, serves as a sharp reminder. The Minister of Planning said as follows: 'Now we have completed our agricultural-development plan. That is fine. What I need now to take to Washington to get financing is projects. So, let us get started.'

PART I

The project cycle, project types and rural development

1 The project cycle

The cycle that projects go through from their initial conceptualisation to the implementation stage – called the project cycle – has been well treated by Baum.[1] An understanding of this process would enable the reader to grasp the intricacies of bringing an investment project to the stage where it is ready for submission to an international financing institution.

The main recognised steps in the process are definition of the project concept, identification, preparation, appraisal, negotiations and supervision. These are all discussed in full detail below. While, for ease of discussion, these steps are treated sequentially in separate phases, it would be wrong to give the impression that all projects follow a clearly defined routine in their formulation. In practice, many of the phases overlap. For example, the project identification stage might include the project-concept definition phase. Sometimes identification may be carried so far into the preparation stage, that for all practical purposes, appraisal could be undertaken immediately. Cases are not unknown where appraisal has been carried out almost simultaneously with preparation. However, at each phase, particular functions are involved which are discussed below.

Project-concept definition

This stage is possibly the most crucial in the project cycle, since 'project ideas' are first generated at this point. In practice, while this stage may frequently merge imperceptibly into the identification stage, it is important conceptually to separate the two stages; because the project-concept definition is prerequisite to the decision to proceed with the next step of identification.

The project-concept definition essentially aims to express the country's development objectives in the form of projects. Ideally, this stage should be based on a thorough understanding of the country's agricultural development objectives, its resource base, and an assessment of the options facing the country. The type of questions that arise could include: should the country invest heavily in irrigated agriculture or should it develop first its potential for rainfed agriculture? Obviously, trade-offs could occur, but the basis for evaluating an appropriate investment blend between the two options should be known. The answers to these questions can be determined reliably only through a sector survey covering both irrigated and rainfed agriculture, where the potentials for development, constraints identified along a time dimension, and relative costs and benefits of main investment options are assessed.

The priority choices lie not only within a particular sub-sector but between sub-sectors. What priority should be attached to livestock development *vis-à-vis* crop development, or fisheries, or forestry? Apart from straight production, what priority should be attached to institutional and support services, e.g. agricultural credit, research, extension, marketing and storage, and what should be the basis for allocation of resources to these ends?

Unfortunately, the genesis of most agricultural projects reaching donor countries and financing institutions does not lie in such a rational process of selection and screening, based on a thorough study of options. The World Bank, recognising this deficiency, undertakes agricultural sector surveys in a limited number of the more important countries. These sector surveys, by focusing entirely on the agricultural sector, provide a more reasoned basis for the assessment of priorities and developing project ideas for the later identification. Possibly, the largest single contribution made by the sector survey to project work is that it examines the role of the project against the background of the country's agricultural development objectives and the way they are being pursued. While the World Bank also carried out economic reviews of selected countries, which do generate agricultural project possibilities, the depth of the economic analysis of the agricultural sector is not as exhaustive as in the case of the agricultural sector study. FAO has also conducted Country Perspective Studies in a few selected countries. These are largely macro-oriented with a long-term perspective, and have not so far been geared to the identification of specific project possibilities. The United States Agency for International Development also does sector work. The IFAD undertakes special programming missions designed to develop an investment strategy that conforms with its lending policies and criteria.

It is not uncommon, therefore, to find that many project ideas are generated on an *ad hoc* basis, and may originate from widely different sources. Borrower countries, consultants, engineering firms, officials of donor countries and bilateral agencies on brief visits to a country are all potential sources. The fact that a particular resource or area has been studied, considerably enhances prospects for its consideration as a project for potential financing. This, coupled with the fact that lending institutions and donor countries have their own aid targets to meet, provides the impetus for project ideas to be followed through to the later stages of the project cycle.

The following example shows the process of developing the project concept in a rural development project in Turkey. The government's proposal was that a broad attack on agricultural problems in the particular province should be started by increasing irrigation, fodder production, feed-lots for cattle, the introduction of imported livestock and the mining of liquite for fuel, as well as strengthening the provincial, social and economic infrastructure, including roads, health services, etc.

A team from the FAO/World Bank Cooperative Programme first examined prospects for making a major impact on agricultural output and farm income especially among the smaller landholders. The conclusion was that there existed good potential for increasing rainfed crop output with minimum long-term investment and that this should be the main approach in formulating the project. There was a demonstrated potential for doubling the yields of dry-land wheat with a foreign variety. Further crop increases could result from improved crop rotation practices, involving reduced fallow land, and higher levels of fertiliser applications, pest and weed control. The increased feed that would become available could increase livestock yields significantly. Irrigated sugar beets, currently produced in the province, could be rotated with alfalfa and other legumes thereby permitting an intensification of cattle production.

The project's target group would be the small farmers in the mountainous areas rather than the medium to larger farmers who cultivated the plains. The former areas suffered from poverty, lack of facilities and technical assistance, and isolation. The assistance to small farmers would take the form of short-term credit for on-farm development, such as forage and fruit-tree establishment; purchases of improved livestock, farm machinery, improved cattle barns and sheep sheds, and hay storage facilities.

Since the majority of villages would require a minimal level of services, a first step would be to develop a number of access roads to those villages currently isolated from main commercial centres, and with some potential for agricultural development. The extension staff would have to be able to service these villages and would require transport and housing facilities. Their numbers would have to be strengthened. The problem of overgrazed communal pastures would have to be tackled through pilot schemes that would examine not only the technical but the sociological aspects of the problem.

In briefly examining the above project concept, the main elements can be stated as follows: first, that the project should be largely based on the development of the area's potential for dry-land farming instead of on irrigated agriculture; second, that the project's beneficiaries should be the small farmers rather than the large farmers, with consequent implications for the location and composition of project activities, and thirdly, that the project should include measures to overcome the more urgent constraints to the area's development, such as improved market communications through access roads, and learning how to solve the problems of institutional practices that contribute to overgrazing of communal lands.

Project identification

The project identification stage develops the project idea to the point where a decision can be taken on whether resources should be committed for the next stage – that is, the detailed project-preparation stage which can be a lengthy and costly process where engineering feasibility studies may be entailed. Whereas the thrust of the project-concept definition is largely the justification of the project in the sectoral context, the thrust of identification is to delineate the main outlines of the project and to establish the overall viability of the project proposal.

Overall viability implies the ability of the project to meet the standards of government and the lending institution. The project's output must have a market. It must be technically, financially and economically viable. Arrangements for the project's organisation and management must be realistic, to ensure the project's success. These project dimensions are thoroughly discussed in Part III of the text.

The major steps in the identification process consist of the following:

1 Evaluation of the present situation
This diagnostic phase has the objective of establishing the development potential for the area or project proposal and identifies the constraints to the proposed development. This is a broad-ranging analysis covering all important features liable to impinge on the project. For a typical project, this would cover the main physical features of the project, soils, agronomic and cropping practices, and water resources; human resources, population and labour; institutional features such as land tenure, and institutions operating in the area concerned with credit, marketing, input supplies, research and extension and the administrative structure.

2 Identifying relevant policy issues
A major requirement at this stage is the examination of relevant government policies and an assessment of their impact on the project proposal. These issues may cover such matters as pricing policies, subsidies, taxation, water charges, etc.

3 Establishing the project's rationale
This provides the overall justification for the country to undertake the project and for the lending institution to support it. A reasoned analysis from the previous sections would provide the basis for marshalling the arguments here. Taking an artesanal fisheries project as an illustration, the diagnosis should have established that the country's fish imports could be substituted with increases in the domestic catch; that fishing is or could become an important economic activity, and that the existing fishery resources are being exploited at levels far below their sustainable yield potential; but evidence should exist that a few entrepreneurs with access to credit, are carrying out a profitable operation.

However, the following constraints would have to be removed if the fish-catch potential were to be realised: the absence of harbour and berthing facilities for fishing boats; inadequate or non-existent shore facilities, including chilling facilities, storage, icing for serving the boats and fish transport; the absence of boats, powered and equipped for longer fishing trips in the deeper coastal waters; lack of foreign exchange for obtaining required fishing gear and replacement spares; disorganised marketing, and the fishermen inadequately trained in improved fishing practices. The project's actions would, therefore, aim to remove these constraints, and develop new economic opportunities in the project area.

4 Developing the project's design and concept
This step consists of setting the project's objectives and the measures proposed by the project to achieve these. This is the core of the project, comprising the project's main components which would have to be financed under the project. Additionally, the project's components may involve measures which require no financing or only very minimal

financing, e.g. the setting up of a project authority, or introducing legislation as required for the project's success.

5 Setting the project's scale
Identification should establish the rough order of magnitude of the project, in respect of physical coordinates, such as the size of the irrigation command area and main project works involved; area to be cleared and planted and families settled in a land-settlement project; number of farmers to be reached and crop rotations to be developed in a credit project. The main factors to consider in the setting of project scale are discussed in Chapter 5.

6 Preparing preliminary cost estimates and benefits
Identification should establish rough estimates of the project's total costs, and of the major project components, and include estimates of the foreign exchange component. The latter is an important consideration in decisions about the size of the loan. Project benefits should also be estimated so as to demonstrate probable economic viability, as there is no point in proceeding with further preparation work if identification establishes that the project proposal has little chance of being viable.

7 Proposing the organisation and management structure
The project's agency could be located either within the main stream of a government department or require special arrangements, for instance, where several participating agencies may be involved requiring a coordinating mechanism. Finalising these arrangements may involve decisions at the uppermost levels of government and be a long drawn-out process. The main outlines of the proposed organisation and management should be sketched out at identification, to form the basis for discussion between the lending agency and the government.

8 Spelling out the further work-requirements
At identification, the studies and other requirements for the detailed project-preparation work should be spelled out, and specific responsibility assigned for various tasks. For example, detailed terms of reference should be drawn up, and estimates prepared of the cost of additional studies and the time schedule.

An illustrative example is a rural development project in a Latin American country, carried out by the FAO/World Bank Cooperative Programme.

Objectives The proposed project is designed with the following objectives:

1 To generate sustainable improvements in living standards among the lowest income groups in a specific region.
2 To increase the production of the agricultural and agro-industrial sectors.
3 To identify the further development potential and needs of the poor zones of adjacent regions.
4 To test and demonstrate the capacity of regional governments to manage complex rural-development projects.

Strategy The project strategy is based on the hypothesis that an impact on living standards can best be achieved by complementing measures designed to generate higher (and more equitably distributed) incomes with programmes to reduce the worst deficiencies in basic infrastructure and services.

Components This strategy would be articulated through a project with the following main components: farm development, irrigation rehabilitation, agro-industries, upgrading roads; improved domestic water systems; electrification; housing; improved health and education facilities; studies for future project formulation; and administration and evaluation.

The identification stage is the crucial one at which various alternatives must be explored. Even if objectives are agreed upon, several options may be open, depending upon the technical possibilities. Irrigation development could entail surface irrigation, and/or groundwater development. A livestock development project could entail substantial imports of breeding stock with attendant risks or a slower build-up from domestic herds. These decisions have implications not only for the project's costs and possible phasing, or development period, but also determine the nature and workload at the project's detailed preparation stage.

In summary, a good identification should accomplish the following:

1 justify the project in the sectoral context;
2 highlight pertinent issues and propose solutions;
3 demonstrate that the alternative or alternatives proposed represent the most efficient use of resources;
4 demonstrate the probable viability of the project proposal and that further resources should be devoted to preparation;
5 establish clearly the follow-up steps for full preparation.

Identifying a rural-development project – an illustration

This project is located in the northern province of a Central African country. The evaluation of the existing situation showed the following features: the area has a population of about 1·7 million with three agro-climatic zones. Smallholders with farm sizes of 2–3 ha predominate, with two-thirds of the cultivated area in millet and sorghum. Cotton and groundnuts are important cash crops. About 2·0 million cattle were raised mainly under nomadic and transhumant pastoralism.

Agricultural development in the region followed three largely uncoordinated approaches: province-wide agricultural support services, settlement schemes to support migration activities and some limited irrigation development. Projects being undertaken by various multilateral and bilateral donors in the province were examined. It was concluded, however, that all these projects not only lacked a properly researched technical base but were being planned and executed outside any overall development strategy, and were affecting only a very small proportion of the population.

Any project proposals, therefore, would have to conform to a development strategy aimed at ensuring coordinated actions; strains in the existing natural balances in the area should be avoided and duplication of services should be reduced to economise on resources, especially manpower. Thus, intensity of support services should reflect population density and agricultural potential of the region.

Based on this strategy the *project's objectives* proposed were as follows: an increase in food and cash-crop production and raising farm incomes of the poorer farmers; improvements in the institutional structure, particularly in the overall planning, evaluation and monitoring of development, and the agricultural support services for both rainfed and irrigated farming; provision of the necessary rural infrastructure for getting services in and produce out of the area; development of a forestry programme as part of an integrated land-use plan for the province.

The project would therefore include the following components:

1 Basic agricultural services package

(a) Consolidation of the existing crop-inputs delivery system, completing the network of stores and points of sale and ensuring the availability of adequate vehicles and staffing for provision and distribution.

(b) Agricultural feeder roads to provide a minimum standard across the province and access to the main producing areas; equipment and technical assistance would be provided for this and for improvement of water supplies.

(c) Rationalisation of the supporting extension and credit services providing additional training facilities, vehicles and equipment.

(d) Assistance with procurement of incremental farm inputs.

2 Development of farm systems and an integrated land-use programme

(a) Further actions to encourage migration from densely populated areas and resettlement south of the Benoue.

(b) Feasibility studies for irrigation development and initiation of pilot projects.

(c) Development of commercial ranching on a pilot scale in the south.

(d) A forestry component to develop further fuelwood and timber supplies under an integrated land-use approach.

3 Complementary institutional and support developments

(a) An applied agricultural research programme catering for the continuing needs of farmers, the project providing necessary buildings, equipment and technical assistance for strengthening the Institute for Agronomic and Forest Research and establishing experimental husbandry farms in the main eco-climatic zones.

(b) Establishment of a competent regional planning, coordination and project monitoring and evaluation organisation, the project providing necessary vehicles, equipment and technical assistance.

(c) Extension of the functional literacy programme.

(d) Support to rural health services if necessary.

As the project was at an early stage of identification, the above list of components should be regarded as tentative. Previous project actions in the province have created at least part of the infrastructure required for provision of basic agricultural services while, institutionally, there are organisations which could provide the basis for the suggested reordering. Consequently, until a full inventory of the facilities in the area is available, it will not be possible to estimate project costs.

Project organisation and management The basis for reordering the institutional structure in the province would be the creation of a provincial development board for overseeing rural development in the province. This board would not be charged with project execution but would take responsibility for planning, coordination, monitoring and evaluation of rural development in the province. Such a board could be based on one or other existing institutions, and would comprise:

1 An executive board to fix development objectives and oversee implementation.
2 An advisory committee to coordinate the activities of executing agencies, to provide a two-way feedback between execution and research and advise on the research programme.
3 A technical secretariat, which in conjunction with other agencies would plan, monitor and evaluate development, execute the study component and coordinate funding in the province. Execution of development actions would remain in the hands of existing agencies.

Project preparation

The identification stage should have established whether there is sufficient justification to proceed to the next stage of project preparation. This can be a costly and time-consuming process, particularly for more complex projects involving large elements of civil works, such as irrigation and drainage. The end result of preparation is a feasibility study which forms the basis on which a financing institution or donor and/or the government can decide about going ahead to appraise the project.

What is the nature and content of such a feasibility study? Naturally this will vary according to the type of project, e.g. irrigation, livestock or fisheries. A summary is given below of the coverage of a rural development project in a province in Turkey, and the format and content of the report.

An introductory chapter briefly sets out government priorities for the region, experience of other rural-development projects in Turkey and previous studies leading to the formulation of this project.

A background chapter should generally cover the country's economic performance including the trade situation; the main characteristics of the agricultural sector including constraints to development; the government's development and social objectives and agricultural policies relevant to the project under consideration, such as subsidies, pricing of inputs, attitudes towards small farmers; and institutions concerned with development in the sub-sector relevant to the project. In the Turkish case, the description of the current economic situation covered the economic growth rate, inflation, employment and trade. The importance of the agricultural sector in the economy was demonstrated, and income distribution and poverty features were highlighted.

A chapter on rural development strategy and project concept establishes the project's rationale and provides the overall justification for the priority leading to the selection of the project. The particular development strategy selected should be justified and there should be some assessment of the risks involved in undertaking the project and possible measures to minimise the risk. Such a risk may arise for example, from uncertainties about the government's ability to meet the estimated local cost requirements of the project, or farmer's response to the formulation of cooperatives proposed for the project area. In the Turkish project, the concept of which was discussed in the preceding section, one such risk stemmed from the group approach, e.g. communal grasslands used in a previous livestock project, and in this project more emphasis was placed on an individual approach to farm development.

A chapter on the project attempts to define the total environment within which the project would have to operate, and provides some assessment of the development potential and constraints in the project area. In this project, the following areas were covered.

Physical features These include the location and climate, topography, geology and soils, and water resources and drainage. The treatment of this topic should be not only descriptive but also analytical. For example, the quantity and distribution of annual rainfall, and the reliability, form the starting point for the justification for an irrigation-development project in the area. Data on geology, topography and soils are required to determine the suitability of the area for irrigation.

Economic base While this section should pay special attention to the area's agricultural base, it is important that the non-agricultural activities receive due attention. This is especially the case where project objectives may include special attention to employment generation. Furthermore, the project area may be remote from the country's main areas of consumption, and increases in production could well lead to falling prices, in the absence of specific measures to generate increased demand.

Included in this section is the area's land-use, e.g. pastures and meadows, cultivated land broken down by main types, forests and other lands. Available data covering field and tree crop production, and livestock production are shown.

Social features These cover such aspects as land tenure, communal pastures, population and migration, labour supply and employment and income levels. With respect to land tenure about one-third of the farm families in the province were estimated to be landless and the larger bulk of them worked on the large holdings. The distribution of farm size was as shown in Table 1.1

TABLE 1.1 Turkey – Farm size distribution in project area

Farm size	Per cent of area	Per cent of farms
Less than 5 ha	26·0	69·3
5–10 ha	25·7	19·6
10–20 ha	20·3	7·7
Over 20 ha	28·0	3·4
	100·0	100·0

Infrastructure The treatment might cover such aspects as the existing irrigation network, roads and railways, and other communications in so far as they are related to the supply of farm inputs and marketing of products from the project area. In the Turkish project about one-half of the 107 000 ha of irrigated farm land benefited from permanent and reliable water as follows.

Permanent irrigation	50 000 ha
surface water modern irrigation	20 000 ha
traditional irrigation (diversion weirs and earth canals)	19 000 ha
groundwater	11 000 ha
Temporary traditional irrigation (mainly in spring)	57 000 ha

Input supply and produce marketing should receive special attention. Distribution systems, including physical facilities for fertilisers, pesticides, etc., should be thoroughly discussed, with special reference to timely availability and the terms and conditions under which farmers receive them. The impact of government agricultural policies with respect to pricing and input subsidies, should be assessed in so far as they relate to actions contemplated under the project.

Government and private institutional structures The discussion and assessment of performance of these in the project area should cover their functions, organisation, staffing, financial resources and their interrelationships. Usually, it is important to identify the extent of centralisation or decentralisation of the decision-making process, as this can be highly significant for the speed and efficiency of project implementation. The purpose of this analysis is to establish the basis for formulating proposals for the project's organisation and management structure.

Subject matter covered in this analysis may include, for example, agricultural administration, irrigation and water management, supporting services such as agricultural research, extension, role of cooperatives, and agricultural credit.

In the Turkish example, the main credit institution is the Agricultural Bank of Turkey which has branches in all districts of the province. A total of about 12 000 farmers was serviced. The major part of the loans are short-term, the demand for medium and long-term credit being constrained by the Bank's collateral requirements which are particularly difficult to meet for the small farmers. The second major source of credit, all short-term, in the province is provided by the Turkish Sugar Corporation to beet growers and livestock fatteners numbering about 8 000 farmers. The analysis concluded that the problem as regards credit institutions in the province is that the main institution because of collateral requirements, lack of specialised staff, etc., is not effectively reaching the smaller farmers but concentrates on the larger, more progressive farmers of the central plains.

All the foregoing analyses form the basis for the chapter on the project. A first section describes briefly the project in general terms, in the second each major project component is discussed in detail and supplementary material provided as may be required. The Turkish project is discussed below, and includes a detailed description of the proposed farm development.

Description

The proposed project has three principal and complementary components.

Production Involves an increased availability of short and medium-term credit to furnish additional resources to approximately 15 000 small and 450 medium-size farms and the development of new irrigated areas, of about 2 500 ha. In spite of the short cropping season irrigation is economically justified by higher yields and wider possibilities for crop diversification.

Infrastructure Involves the construction of access roads to link currently isolated groups of villages that have a good potential for development, to commercial centres and the provision of basic village facilities and social services (schools, health centres, drinking water, technical staff, housing, input stores) where justified.

Institutional support With the aim of building-up, reinforcing and, if necessary, equipping existing institutions which would be involved with extension, local adaptive research, credit supervision, irrigation and handicraft (carpets) development.

Detailed features

Farm development The project is specifically directed towards owners of small and medium-size farms and additional credit is the main production component on which the proposed farm development is based. Farmers would use short-term credit to purchase seeds, fertilisers, chemicals, medicines and dips from the government's agricultural input supply institution, and to hire mechanisation services and pay for hired labour in certain cases. Farmer participation, at full development is assumed to reach: 10 000 small farms with 3 ha of cropland (23 per cent of total), 5 000 small farms with 7 ha of cropland (37 per cent of total), 200 medium-size farms with 17 ha of cropland (4 per cent of total; they are assumed to engage in intensive dairying) and 250 farms with 30 ha of cropland (20 per cent of total) requiring respectively an annual maximum of short-term credit of Turkish lire (TL) 5 270; TL18 460; TL55 000 and TL62 600. The total requirement would amount to TL174·35 million and involve an area of 75 000 ha.

Medium-term credit would finance all aspects of farm development, including:

1 3 200 modern bee hives for 800 small farms (3 ha);
2 improvement of cattle barns and sheep sheds, including sheltered hay storage space on 3 050 farms (800 farms of 3 ha, 2 000 farms of 7 ha and 250 farms with 30 ha);
3 purchase of a maximum of 1 000 pure-bred bulls for 2 000 farms (7 ha);
4 purchase of 2 200 motor mowers for 7 ha and 17 ha farms;
5 establishment of 0·5 ha of orchard on 200 farms (7 ha) or a total of 100 ha of orchards;
6 for each of the 200 intensive dairy farms (17 ha):
establishment of 3 ha of alfalfa,
purchase of 4 pure-bred heifers,
construction of a new cattle barn;

7 purchase of 250 tractors with implements for 250 medium-size farms of 30 ha.

The total amount of medium-term credit necessary would reach TL181·3 million.

Project costs and phasing should be based on some realistic estimates about the rate at which the major proposed project actions can be implemented. Thus the phasing should cover not only the physical works to be undertaken by the project, but also other project components, such as agricultural credit, the establishment of a management structure and the undertaking of studies for later phases of the project. This cost phasing sets the basis on which the lending institution derives its disbursement schedule for the loan and therefore should be as realistic as possible, since lagging disbursements suggest that one may be dealing with a problem project.

The table summarising costs, which should be fully documented by supporting material, should show the capital or investment costs, broken down under major headings of investment, and operating and maintenance costs, also called recurrent costs. Table 1.2 provides such an example. These costs should all be expressed as 'financial costs' and not 'economic costs' since the purpose is to establish what will actually be financed or the cash outlays of the project. The cost table should also indicate the breakdown as between foreign exchange and local-cost components. The foreign exchange component usually constitutes one reference point which aids donors to base their decision about loan size. It should be noted that the costs include a provision for physical contingencies. Estimates for price contingencies may or may not be included at the preparation stage but certainly must be included at the appraisal stage.

Organisation and management This chapter should spell out how the project is to be executed, and should allocate responsibilities for the major actions proposed by the project, and propose measures for coordinating the activities of the different responsible agencies or bodies involved in the project. When technical assistance is included to assist in project implementation, provision should be made not only for costs, but for clear definition of duties and responsibilities of experts.

In the Turkish case, the proposed organisation and management consisted of the following main elements:

1 A National Advisory Committee based in Ankara with representatives at under-secretary level of the different ministries involved in the implementation of the project, and a Provincial Advisory Committee based in the province. The latter would include the Governor as Chairman, with the project manager as secretary and with representatives, at director level, of the different agencies involved.
2 The establishment of a qualified management office in the province, headed by a rural-development project manager with responsibility for supervising project actions to be carried out by different agencies.
3 The provincial extension service would be the technical arm for transfer of technology and the preparation of mainly crop-oriented on-farm development plans. It would be reorganised, staffed and equipped accordingly.

TABLE 1.2 Turkey: Rural development project – Total project costs

	(TL '000)								
	Year 1	**Year 2**	**Year 3**	**Year 4**	**Year 5**	**Total**	**Total** (US$ '000)	**Foreign exchange** (per cent)	**Foreign exchange** (US$ '000)
Investment costs									
On-farm investments	18 150	36 000	36 300	45 201	45 309	181 260	10 358	40	4 143
Irrigation	16 000	28 000	36 000	42 000	18 000	140 000	8 000	20	1 600
Rural roads	35 000	35 000	35 000	35 000	35 000	175 000	10 000	40	4 000
Social infrastructure and handicraft activities	300	46 500	45 300	44 400	44 400	180 900	10 337	20	2 067
Project offices equipment	42 300					42 300	2 417	20	483
Research equipment		5 250				5 250	300	40	120
Irrigation study	2 000					2 000	114	10	11
Sub-total investments	113 750	151 050	152 600	166 601	142 709	706 710	41 526	30	12 424
Short-term credit	18 475	35 650	34 350	42 906	42 969	174 350	9 963	20	1 993
Operation and maintenance costs									
Irrigation		160	515	950	1 482	3 107	177		
Rural roads		1 400	2 800	4 200	5 600	14 000	800		
Project offices	12 438	12 438	12 438	12 438	12 438	62 190	3 554		
Training		1 000	1 000	1 000		3 000	171		
Research			1 750	1 750		3 500	200		
Sub-total	12 438	14 998	18 503	20 338	19 520	85 797	4 902	10	490
Total	144 663	201 698	205 453	229 845	205 198	986 857	56 391	26	14 907
10 per cent physical contingencies	14 466	20 170	20 545	22 985	20 520	98 686	5 639		1 491
Grand total	159 129	221 868	225 998	252 830	225 718	1 085 543	62 030		16 398

It should be noted that the costs include a provision for physical contingencies. Estimates for price contingencies may or may not be included at the preparation stage but certainly must be included at the appraisal stage.

Apart from the general aspects, more specific proposals must be made covering important project activities. For example, such areas as agricultural credit, marketing, supply of inputs, research and extension services, cooperatives and role of farmer organisations should be covered.

Production, markets and financial results This chapter attempts to measure the incremental output on project farms arising from the project's investments and establishes that the project's output can be satisfactorily marketed and that the financial results to the intended beneficiaries are such as to generate their participation in the project. This latter is demonstrated through a series of farm models.

These points are briefly illustrated in the Turkish case. The annual incremental production of grain from the project would be 5 400 tonnes of wheat and 300 tonnes of barley, at full development, with a total annual value of TL14·9 million. The increased wheat production represents about 4 per cent of local wheat production and less than 1 per cent of the local barley production. The existing shortfalls of wheat production in the region ensure that there would be no problem in marketing the incremental production.

The incremental production of sugar beets would be around 32 900 tonnes at full development, or about one-third of current production in the province. Total value of incremental production was estimated at TL22·3 million. The increased sugar production from the project would contribute to satisfying national demand which has been growing at annual rate of 6 – 8 per cent over the past ten years.

Lastly, the effects of the project on farm income are shown by farm budgets which translate these plans into financial terms. Examples of these findings are: for a 3-ha farm, with a loan recovery period of 8 years, including a grace period on repayment of 2 years, the net income would rise from TL8 840 without the project, to TL13 640 with the project, and the financial rate of return on the total on-farm investment would be about 35 per cent. In this case, farm models were prepared for farms ranging up to 30 ha.

Project benefits and justification This final chapter should pull together the evidence to demonstrate the benefits to the country of undertaking the project. The single most important criterion used by financing agencies is the 'economic rate of return'. This is defined as the rate of discount at which the total present value of costs incurred during the life of the project is equal to the total present value of benefits accruing during the life of the project. While the minimum acceptable rate of return should be related to the opportunity cost of capital in the particular country, a rule-of-thumb figure is that a 10 per cent economic rate of return provides a good baseline on which to judge the acceptability of the project for financing.

The attainment of other project objectives should be assessed here. The more important ones are the employment generated by the project, both during the construction period and after the project is fully operational. Project effects on the alleviation of rural poverty, on the pattern of income distribution, on nutrition, on improved standards of living deriving from health and water supplies, should be quantified and highlighted where possible. The project's contribution to institutional improvements, such as administrative services, should also be discussed.

Project appraisal and loan negotiations

The appraisal stage consists of an independent check by the lending institution and/or aid donor of the recipient country's submission. While the term appraisal generally implies an activity undertaken by an external financial agency, it is not uncommon for local or domestic agencies to undertake their own appraisal simultaneously within a framework agreed upon with the external financing agency. For example, the external agency may approve a line of credit to a national development agency, which would have to undertake appraisal of sub-projects on the basis of criteria agreed with the former. In a state-wide programme for irrigation development, involving a wide range of small discrete project areas, the agency appointed for project implementation might be assigned responsibility for appraisal of the sub-projects, again, along previously agreed lines. The government itself may also undertake appraisal of a project intended for its own exclusive financing.

While appraisal does in practice cover much the same ground as that of identification and preparation, there are specific areas, mainly involving project financing, that receive special emphasis. Appraisal tries to establish that the major assumptions in project formulation are, in fact, correct and realistic. Technical assumptions relating to yields, cropping patterns, etc. would be checked. In projects involving large civil engineering works, e.g. construction of a large dam, considerable attention is paid to ensuring that required technical standards are fully observed, e.g. the question of safety. The project's cost estimates are checked and so are all the assumptions relating to the economic and financial viability of the project. The project's administrative feasibility is checked by reviewing the proposals for organisation and management at the time of project preparation. Appraisal must establish that the project has a satisfactory economic rate of return.

Appraisal however has some special tasks which are distinctly related to the donor/client relationship. The more important ones relate to the financing package, conditions of disbursement, procurement and determining the issues leading to loan negotiations.

The financing package

The putting together of a financing package can be a relatively simple affair, but can also be quite complicated depending on the nature of the package. An illustrative example of a simple case for one rural development project is as follows: the project's total cost was US$50 million, and the foreign exchange component of US$21·6 million meant a foreign exchange of 43 per cent. The financing decision was as follows.

The proposed loan of US$25 million would finance the full foreign exchange costs and about 12 per cent of the local costs of the project. It would cover 50 per cent of total project costs. The government would provide the remaining US$25 million out of annual budget appropriations. Each implementing agency would maintain a separate account for the project, and would submit all budget requests to the project director for approval. The project director would forward approved requests to the commissioner of the budget, who would disburse all project funds directly to the implementing agencies, in accordance with normal government procedures which are satisfactory to ensure the continuous and timely flow of project funds.

The financing package becomes more difficult to assemble in projects involving co-financing. This is a generic term referring to cases where two or more lending agencies and/or donors join together in a coordinated way to undertake their loan activities and establish conditions for a particular project. Broadly, co-financing takes two forms, namely, joint financing and parallel financing. In joint financing, the funds of the co-lenders are blended together and there is no attempt separately to finance clearly identifiable components. There is a common list of goods and services, and the financing of all or certain items is shared between the co-lenders in agreed proportions. In parallel financing in contrast, the co-lenders deliberately would select certain project components, that is, they would finance different goods and services or parts of the project. For example, in a multi-purpose water development project, one agency might finance the electrical power-generation aspects; a second agency might finance the dam and irrigation network, while a third might finance the downstream activities of land development and agricultural support services.

The type of financing has very direct implications for the procurement practices that the borrower would have to adopt in the acquisition of goods and services for the project. Thus, a multilateral financing institution, as a general rule, would have to employ international competitive bidding open to all its eligible member countries, for those project components that it finances. In contrast, bilateral donors by and large, prefer parallel financing, since this permits them to insist on tied procurement, thereby providing a monopoly to suppliers in the donor country for bidding for those goods and services that the latter has selected for financing. Joint financing as a rule does not allow room for employing different procurement practices by different co-lenders. Apart from procurement considerations, public relations considerations apply as different co-lenders may wish to have an identifiable presence in the country, an opportunity best afforded by parallel financing.

The following case (Table 1.3) illustrates the application of both joint financing and parallel financing in the same project, for an irrigation and drainage project. The institutions are labelled as lending agencies (LA 1 and LA 2), while the bilateral donor is designated as BD 1.

It will be noted that the two institutions LA 1 and LA 2 would finance a joint fund totalling US$20·0 million with roughly a 50:50 per cent split of selected items, except for the item, Advance for project preparation, which was a reimbursable loan previously advanced by LA 1 to finance feasibility studies leading to this project. The package financed

TABLE 1.3 Summary project costs and financing plan (US $ million) for irrigation and drainage project

Item	Total cost	Financing			Government and other co-financers
		LA 1	LA 2	BD 1	
Civil works	15·600	5·160	5·160		5·280
Buildings	0·800			0·800	
Equipment, including pumps	6·361	1·920	1·920	2·160	0·361
Engineering and supervision	0·667				0·667
Consultant services	3·012			2·720	0·292
Extension and research	0·382				0·382
Advance for project preparation	0·280	0·280			
Further feasibility studies	0·400			0·300	0·100
Sub-total	27·502	7·360	7·080	5·980	7·082
Physical contingencies	4·128	0·920	1·200		2·008
Price contingencies	11·204	1·720	1·720	1·520	6·244
Total	42·834	10·000	10·000	7·500	15·334
Percentage	100	23	23	17	37

by these two institutions therefore comprises a joint financing operation with a common list of goods and the blending of funds. The procurement guidelines for LA 1 would govern the acquisition of goods and services financed by the two institutions.

In contrast, the loan from the bilateral donor BD 1 would be used on a parallel financing basis to finance a separate and identifiable package of goods and services and involve tied procurement. Since the rest of the financing package (US$15·3 million or 37 per cent of total project costs) would have been beyond the country's financial capacity to bear, other co-lenders were invited in an effort to close the financing gap of this project.

Briefly, the practice of co-financing has been expanding considerably in recent years. Among the more important reasons are: good, viable projects in agriculture are hard to find and are slow in preparation; in addition, there are many donors in search of such projects.

Many new financing institutions, for example, some in the Middle East, do not have the facilities, including staff, to develop their own project pipelines. They must, therefore, depend largely on established and experienced institutions to find projects. Sometimes the size of individual projects may be so large as to exceed the capacity of a single donor to finance. There is also the important question of country programming by the institutions and bilateral donors, particularly in regard to the use of concessional funds. Thus, an institution or donor might have made specific allocations to various economic sectors, e.g. roads, mining, in a particular country, and its allocation for agriculture may just be inadequate to cope with the size of a proposed agricultural project.

While from the standpoint of the borrowing countries there is no clear-cut indication that this surge in co-financing agricultural projects has led to an increase in resource flows, there are other benefits to them.

First, it broadens the range of their shopping operations for financing, thereby reducing excessive dependence on any one particular lender. Since terms of lending may vary by institution and donor, the borrower may well end up with a softer blend in terms of the overall cost and repayment period of the financing package.

Procurement

At appraisal, the procurement practices covering the acquisition of goods and services are clearly spelt out. These may cover not only the bidding practices to be followed but may stipulate margins of preference. A brief example of this follows for one actual case.

Contracts of US$100 000 or more for purchase of vehicles and equipment (US$3·0 million) would be awarded on the basis of international competitive bidding (ICB). Domestic manufacturers would be granted a preferential margin in bid evaluation equal to the prevailing tariff, or 15 per cent of the c.i.f. cost of imports, whichever is the lower. Civil works contracts totalling US$100 000 or more (US$1·2 million) would be subject to ICB; in evaluating bids, domestic contractors would be allowed a 7·5 per cent preference. Most contracts for building construction (US$5·7 million), which would generally be small, of varied design and geographically scattered, would be constructed by a combination of local competitive bidding, force account and self-help.

Disbursements

For the same project, the loan funds would be disbursed according to the following percentages and amounts:

1 100 per cent of foreign expenditure and 85 per cent of local expenditures for vehicles and equipment (US$3·0 million);
2 75 per cent of total expenditure for operation and maintenance of vehicles and equipment (US$4·1 million);
3 100 per cent of foreign expenditures and 75 per cent of local expenditures for technical assistance, staffing and training (US$5·7 million).

The percentages are indicated for all the major investment categories. At appraisal, the schedule of disbursements should be drawn up, preferably by quarters, during each fiscal year of the disbursement period. The table should also show the cumulative disbursement and the balance of the outstanding loan.

Loan negotiations

The end result of loan negotiations is the drawing up of a set of loan documents, for example, loan agreement, or credit agreement, guarantee agreement, subsidiary loan agreement and/or joint project agreement which are legal instruments with covenants binding on both the lender and the borrower, and the guarantor, where applicable. The latter arises in cases where the government itself may not be the borrower, but the borrower is, for example, a National Development Bank or Development Corporation with powers to borrow abroad. The government may then be required to guarantee the loan repayment.

Several covenants in the loan instrument may derive directly from the findings and recommendations at appraisal. An illustrative example for a land-settlement project in one Asian country is given below.

During negotiations the following principal assurances were given:

1 The National Settlement Corporation would appoint by 31st December, 1974 a deputy general manager, chief agricultural officer and civil engineer and would not, before completion of the project, undertake additional substantial settlement activities without consultation with the lender.
2 Settlers would be selected according to criteria agreed with the lender, and project sites would be protected by notices and effectively guarded to prevent premature settlement.
3 The Settlement Corporation would not give approval for farm credit without a farm plan having been prepared by the Corporation or the Agricultural Development Bank, and accepted by the settler borrower; and the latter would supply appropriate credit to project farmers.
4 The project's general managers would be responsible for monitoring and evaluating project agricultural and forestry operations and would promptly submit project evaluation reports to the lender.

The following would be conditions of effectiveness:

1 That a subsidiary loan agreement between the government (the borrower) and the Agricultural Development Bank had been signed.
2 That the Settlement Coordinating Committee had been established.
3 That the National Settlement Committee had submitted to the lender satisfactory evidence of its right to operate and hold lands for the project in the three sub-project areas.

It should be noted that the loan cannot become effective, that is the loan agreement does not come into force before the borrower has satisfied the lender that the conditions of loan effectiveness have been met. For example, the executive branch of the borrowing government may not be able to sign the formal loan documents until after parliamentary approval of the loan has been obtained.

Negotiations permit the representatives of both the lender and the borrower to defend their respective positions regarding various policy issues that are somehow related to the execution of the project. Any covenants, for example, that tend to be restrictive on the borrower's right to change his administrative machinery as he sees fit, are highly contested, as the borrower may get the impression that the sovereignty of the country is being questioned.

Supervision

The purpose of supervision is to ensure that the project's implementation is carried out to acceptable standards, and that the project's targets are being realised. Whereas in the previous stages of the project cycle, an identity of interests results in a joint responsibility for project formulation between the aid donor and the borrower, after the loan resources become effective, the execution of the project becomes the prime responsibility of the borrower, whereas supervision is the sole responsibility of the aid agency.

The supervision process entails both a review of the progress reports required of the borrower, and field visits to the country, with problem projects requiring more frequent supervision. To date, the supervision function has operated largely as an external audit, where such items as rate of disbursement of loan funds, procurement practices, and organisation and management features received great attention. Problem areas are identified and reports of the supervision teams form the basis on which the borrower may be requested by the aid agency to take certain corrective measures.

An important spin-off of the supervision process is that it sometimes generates *repeater projects*, which are really follow-up phases of previously financed projects. These are quite common in agricultural credit projects, where a first project may have the major objective of building up a credit institution so as to be able to undertake larger loan operations. A previously untested technology, for example, a new pasture grass could, after its satisfactory performance, lead to further livestock projects. Finally, at the end of the implementation, there is usually a *completion audit* to assess overall performance.

Technical assistance in project preparation

The virtual absence of the capacity for project preparation in developing countries is accepted as a major constraint to the rate at which aid-flows to agriculture can be accelerated, without unduly affecting the quality of projects financed. General agricultural planners, frequently met in many countries, lack the basic experience required to formulate projects in agriculture to standards acceptable to the lending institutions.

In recognition of this, the main lending institutions, spearheaded by the World Bank, found it necessary to formalise arrangements for extending technical assistance in project formulation to the developing countries. In 1964, FAO and the World Bank, under a cooperative agreement, established a team of agricultural specialists, with the primary responsibility of assisting governments to formulate project requests in agriculture for submission for financing. The main principle, guiding the work of the cooperative programme, is that every project has several facets to it and requires a multidisciplinary approach in its formulation. A further consideration is that the interests and objectives of the lending institution may not always coincide neatly with those of the borrower country, and that an independent objective viewpoint could be advantageous to both parties. It is not surprising, therefore, that this programme now prepares projects for countries, accounting for over 35 per cent of the World Bank's lending for agriculture. Additionally FAO has developed similar working arrangements with the main regional lending institutions.

Apart from this, there is a whole range of consulting firms, mainly based in the developed countries, who are equipped to assist countries both in the conduct of feasibility studies required during project preparation and also in project implementation. Such firms are especially competent in the preparation of feasibility studies involving civil engineering works but are generally less competent in organising the human elements of projects, e.g. farm-level organisations. Within recent times, the costs of using consulting firms has risen very substantially, and many countries are just not able to finance the cost of such feasibility studies. Accordingly, the main multilateral lending agencies all have some provision for financing the technical assistance required for project preparation. In some cases, a non-reimbursable loan may be given for the purpose. However, in cases where such a technical assistance loan leads later to an investment project for which the institution makes a loan, then the country may be required to repay the costs of the earlier technical assistance loan.

References

1 Warren C. Baum, The Project Cycle, in *Finance and Development*, Quarterly No. 2, 1970 (a publication of the International Monetary Fund and the World Bank Group). An updated version of this by the same author appears in *Finance and Development*, Vol. 15, No. 4, 1978.

2 Definition of and reasons for projects

This chapter reviews the more widely accepted definitions of a project, and discusses the reasons for undertaking projects.

Definition

A first distinction must be made between public and private sector projects. The former, which are the main targets of aid-flows discussed in the Introduction have their origin in government plans for the agricultural sector, and make demands on governmental resources and institutions, even though the private sector may be involved in the project's implementation. While private-sector projects may be important in countries with a large estate sector, and may involve external private-resource flows, these fall outside the range of this study.

A second distinction is frequently made between investment projects and pre-investment projects. While objectives may differ, they are complementary, because pre-investment projects can and should set the basis for investment decisions. The UNDP pre-investment activities comprise mainly resource surveys, feasibility studies and institution-building.

The main differentiating feature between agricultural projects and projects in other economic sectors is the biological nature of agricultural pursuits, making it harder to predict input–output relationships. In many developing countries, agriculture is a way of life which should be fiercely protected from intrusions.

Examples of definitions of a project include Gittinger's definition:

> Generally, in agricultural projects we are thinking of an investment asset from which we can expect to realize benefits over an extended period of time.[1]

FAO[2] thinks of a production project as a proposal for investment with the definite aim of producing a flow of output over a specified period of time. A production project always involves a stream of costs, that is the engagement of physical and human resources for its realisation and flow of future benefits in physical terms.

Lewis[3] provides a different perspective. The administrator of non-political aid finds that choosing projects is not as satisfying as choosing people. Projects are words and drawings and figures on paper, and interpreting them requires expert knowledge that the administrator does not possess. King[4] puts it simply by calling a project a 'proposal for a capital investment to develop facilities to provide goods and services'.

The foregoing definitions of a project all tie investment activity to benefits. The emphasis is on resource flows and output streams, suggesting an almost passive relationship.

In contrast, we see a project in a much more dynamic context, and propose this new definition:

> A project is an instrument of change. It is a coordinated series of actions resulting from a policy decision to change resource combinations and levels so as to contribute to the realisation of the country's development objectives.

This definition forces the project planner first to set the project's objectives within the framework of the country's development priorities and second, to ask probing questions about what is to be changed, the reasons for change, and the strategy for implementing change. Development objectives of the country may cover such aspects as expanding the agricultural production growth rate, savings or generating foreign exchange, reducing unemployment, improving income distribution, eradicating poverty and malnutrition, improving the lot of specific ethnic groups, and promoting larger public involvement in the production of goods and services.

Reasons for undertaking projects

A project, being an instrument of change, is undertaken to change the dynamics in a system that is at an undesirably low level of equilibrium or that is stagnating. In all cases, the primary objective is to get people to act in a predetermined way. A project is thus people-oriented in the first instance, with the ultimate purpose of evoking response to a particular action or a group of actions.

The project analyst, therefore, tries to forecast the reactions or response level to his project actions. Some examples will illustrate this. In a seeds project, the project actions could well result in producing a seed stream of superior quality to that currently used by the farmers. It is, however, the farmers' response to this improved seed stream that dictates the level of influence that the project would have on increased production at the farm level. The human element comes into play when the farmer carries out his own cost–benefit analysis of the technical and economic merits of the plant breeder's work, which now becomes a new input in the farmer's investment decision-making.

Irrigation projects always involve some elements of forecasting farmers' response, since the benefits stemming from the project's investments are sharply determined by the farmers' decision to make the derivative and complementary investments in land preparation and crop production, and by the rate at which he does them.

However, while the most frequently encountered projects are those in which public policy aims at *inducing* the farming population to undertake certain production objectives by committing their own resources, there are also projects in which government intervenes directly in the production process itself, with its own resources.

National priorities and targets

Most developing countries now have some sort of a national development plan – usually covering four to six years. The plan establishes not only the overall growth-rate target for the planning period but breaks this down by sectors. Investment budgets are drawn up and certain assumptions are made about incremental capital/output ratios, so as to arrive at the investment requirements for achieving the desired growth rate.

The overall plan expresses the country's broadest objectives at the macro-economic level. Derivative plans result in the development of programmes which express themselves at the micro-level in the form of specific projects. We can, therefore, think of projects as the final link in the planning chain or basic building blocks. The extent to which these projects are conceived and implemented in fact determines whether the global objectives of the plan will be realised. The project is thus the point at which the plan becomes action-oriented.

The differentiation between a programme and a project is important, especially from a financing standpoint. Although one can conceive of overlapping situations, a programme has generally a much broader range of objectives and implications than a project. A programme tends to be national in scope, and reflects the totality of the government's objectives for a particular sub-sector. In contrast, a project tends to be regional or location-specific. Thus, a livestock-development programme could include, for example, animal health services; grassland and pasture research; livestock breeding and testing of imported breeds; livestock production, covering large and small livestock; marketing and processing of livestock products and regulatory activities. A livestock project, on the other hand, could include all the above components but the focus, the scale of overall effort, and the allocation of investment to various components would be sharply different.

At the project level, components are only introduced as they contribute to the realisation of the objectives set for the project. To illustrate this, a dairy project to meet the milk requirements of a specific urban area, might initially concentrate on financing the investments required to raise milk production in the areas normally supplying the city, where an established demand exists. A dairy-production programme for the whole country might, however, require special measures to introduce and develop milk-drinking habits among the population, thereby creating a larger demand for fluid milk.

In a comprehensive study on development planning, Waterston[5] refers to the widespread situation in which budgeted development funds remain largely unspent during the plan period. He attributes this situation to the lack of absorptive capacity characteristic of the less developed countries. This essentially reflects an inability to invest in soundly conceived development programmes and projects that can be carried out well and operated successfully upon completion. Many reasons account for this, such as lack of trained staff for preparing and executing projects, poor phasing of projects and laxity in meeting time schedules, inflated budgetary estimates, cost over-runs, poor construction, etc. Waterston also refers to the fact that because some governments consider investment virtually synonymous with development, they tend to emphasise the fulfillment of the financial-investment targets in their plans rather than the physical-output targets which the investments are aimed at achieving, thus confusing the means (investment) with the objective (output).

Yet, all this tends to suggest that most countries are better equipped at preparing national development plans, than preparing specific projects. In fact, the sectoral projects in the plan to a large degree reflect the differential capacity among various ministries and agencies for preparing projects. Waterston cites the case of Morocco, where the Ministry of Public Works spent 75–80 per cent of the funds made available to it, while the Ministry of Agriculture spent as little as 30 per cent of its annual budgetary investment allotments. One other factor at work could be that the planner at the national level tends to underestimate the time required not only for the preparation of feasibility studies of projects to the standards acceptable to international lending agencies, but also underestimates the severity of the problems involved in the project's implementation.

We now narrow the discussion to agricultural plans, programmes and projects. The literature abounds with books, reports and publications on this subject. Many international agencies, such as FAO, the World Bank, the regional banks for Latin America, Asia and Africa, and the Organisation of American States, provide assistance to countries in either reviewing or developing agricultural plans.

It it useful to examine briefly one such plan for Nigeria's Agricultural Development, as prepared by FAO.[6] The major planning premise from which the agriculture plan was developed was that the economic and social objectives of development are matters to be determined by the people themselves, through their elected representatives, and that FAO considered that it has a responsibility to work within the framework of these objectives which derive from a set of political decisions. Against this background, the major tasks, facing Nigerian agriculture were identified as the provision of the following:

1 an adequate and well-balanced food supply for the growing population;
2 agricultural raw materials for Nigeria's developing industries;
3 the volume of exports needed to pay for the import of capital goods;
4 employment for the additional agricultural working population;
5 a substantial share of the capital to finance the development of the whole economy.[7]

The bulk of the report then deals with policies and programmes for various major crops, cacao, rubber, oil palm, forestry, livestock, processing industries, and also institutional matters, such as land tenure and organisation of services. The main oil palm programmes recommended were a large-scale rehabilitation programme of some 238 760 ha over 12 years in eastern Nigeria and a new planting programme for smallholders of some 48 562 ha. For western Nigeria, new plantings of 80 937 ha were recommended. Financial requirements were roughly worked out. However, the last link of formulating precise projects to acceptable standards was done after completion of the plan.

At this stage, it is important to recognise that the priorities at the national level may not coincide at all with those of the farming sector, particularly where government policy may deliberately foster terms of trade prejudicial to the agriculture sector, favouring, say, the manufacturing or consumer sectors. A government, experiencing severe foreign exchange problems and requiring capital-goods imports for the country's development, would naturally give priority to those projects with large foreign exchange benefits. Promotion of industrial crops, such as rubber, oil palm or jute, could then command a higher allocation of public resources than would, for example, food crops with a primarily domestic end-use, especially where little or no important substitution is involved. For the larger bulk of the farming population, lying within the subsistence sector, the immediate priority is ensuring the family food supplies, with the profit motive being a secondary consideration. A programme or project aimed at self-sufficiency, in, say, rice production, is liable to strike a more sympathetic note with the small farmer since his immediate objective harmonises more closely with the national objective.

While national increases in agricultural production constitute by far the major reason why countries undertake projects, and foreign exchange considerations loom large in this, other considerations come into play.

Thus, many projects may aim at bringing unutilised or under-utilised resources into production. Transmigration projects in Indonesia aimed at bringing the land resources of the outer islands into production and thus relieving the pressure on Java are a case in point. The harnessing of water resources for electricity and irrigation is quite common. The Olmos project proposal in Peru has the objective of tapping the water resources east of the Andes mainly for irrigation of the dry coastal plains of the country, through the construction of a tunnel and an irrigation distribution network.

Projects may also be necessary to pursue agricultural diversification policies in countries with an economy dominated by one or two major crops. Bananas and coffee in many Latin American countries are cases in point. The sharp swings in export earnings make the countries cyclically vulnerable, and diversification of the crop-mix contributes to greater stability of earnings.

The policy framework of projects

Projects do not operate in a vacuum; their performance is subject to a wide range of government policies covering such matters as agricultural taxation, subsidies, land reform and tenure, level of priority of foreign exchange allocations to various economic sectors, and pricing policy and agricultural credit.

Practically every agricultural production project requires the project planner to make some assumptions about future price levels of products being financed so as to assess the project's financial and economic viability. Thus, the role of pricing policy, as it relates to projects, warrants special attention. A judgement about the impact of pricing policy is important to assess the probable rate of uptake of loan-funds by potential project participants. In a livestock-development project, for example, the rate of uptake of loans for the development of beef enterprises will almost certainly be affected by farmers' perception of how government-controlled meat prices would affect the profitability of their operations. A local on-lending credit institution, employing funds from external aid agencies, would certainly have to reckon with the pricing policy in trying to assess effective demand for loan funds. In one Caribbean country, the relaxation of controlled beef prices led to an immediate and rapid uptake of loans for beef enterprises from the credit institution.

The pricing effect cannot, however, be seen in a very narrow context. The literature is replete with discussions on the role of pricing policy in agricultural development. Mellor[8] discusses the limitations of price policy in increasing the aggregate of production in traditional agriculture systems, largely because of inconsistencies with other basic development goals, and the problems of organising and administering appropriate devices for effecting price increases.

Uma Lele,[9] in examining the impact of pricing and marketing policies on rural-development projects in Africa, concluded that these policies, especially towards food crops, exerted a substantial adverse effect on the performance of such projects, and that the pricing element was a factor conditioning the rate of adoption of new technologies in the programme areas.

Krishna[10] in a comprehensive review of price policy concludes that available evidence indicates that wherever a minimum development of transport and monetisation has been completed, the output of individual crops as well as aggregate output responds positively, at the margin, to favourable relative price movements, subject to constraints of resource availability. However, there is little reliable evidence available concerning the response of the marketed output of subsistence crops to the relevant relative price movements, and in some cases the response could well be negative. Krishna further concludes that the growth of agricultural output has to be induced mainly through institutional and technological improvements, and a great increase in the supply of inputs required by these improvements. The role of price movements is either to accelerate or retard, or to arrest these changes.

The implications of these findings for the project analyst are that price can only be viewed as one contributing element in a package intended to provide an incentive to undertake the cultivation of specific crops or the technological changes induced by the project. The question of incentives is discussed in Chapter 5. The response to prices is liable to be larger in cash crops, grown for export or the domestic market. In projects oriented towards small farmers and the rural poor, the impact of price on aggregate marketed production of food crops is apt to be small or negligible, because of the many constraints the small farmer faces, and because additional production is channelled first to increased family consumption.

In assessing the impact of price at the farm level, the project analyst must be guided, not so much by the guaranteed minimum price, when this exists, or market prices at various points of the chain, but by the producer price – that is, the farm-gate price. One useful way to gauge the impact of product price on production is to relate this to the price of purchased farm inputs. What does a tonne of urea fertiliser cost, and what additional production does it generate in income terms? Thus producer price is the starting point to assess whether a particular recommended package makes sense at the farm level.

Land-tenure systems are discussed in Chapter 8, as a component of the overall institutional framework, and how they relate to projects.

References

1 J. Price Gittinger, *Economic Analysis of Agricultural Projects*, the Economic Development Institute, International Bank for Reconstruction and Development, Johns Hopkins University Press, Baltimore, 1972.

2 FAO, *General Guidelines to the Analysis of Agricultural Production Projects*, Agricultural Planning Study No. 14, 1971.

3 W. Arthur Lewis, *Development Planning*, New York, 1966.

4 John A. King (Jr), *Economic Development Projects and their Appraisal*, the Economic Development Institute, Johns Hopkins University Press, Baltimore, 1966.

5 Albert Waterston, *Development Planning Lessons of Experience*, The Economic Development Institute, International Bank for Reconstruction and Development, Johns Hopkins Press, Baltimore, 1965, pp. 300–304.

6 FAO, *Agricultural Development in Nigeria, 1965–80*, Rome, 1966.

7 *Note:* this policy objective would now be outdated with Nigeria's development of its oil resources.

8 John W. Mellor, *The Economics of Agricultural Development*, Cornell University Press, Ithaca, New York, 1966, p. 206.

9 Uma Lele, *The Design of Rural Development, Lessons from Africa*, a World Bank Research Publication, Johns Hopkins University Press, Baltimore, 1975, p. 181.

10 Raj Krishna, Agricultural Price Policy and Economic Development, in *Agricultural Development and Economic Growth*, Herman M. Southworth and Bruce F. Johnston, (eds) Cornell University Press, Ithaca, New York, 1967, pp. 497–540.

3 Characteristics of main project classes

The definition of a project, earlier proposed, focuses on 'change' as the primary purpose for undertaking projects. We should, therefore, be able to develop a grouping for projects, based on the objectives of change. A proposed functional classification would be:

technological innovation
broadening the physical resource base
improved status of disadvantaged groups
improved post-harvest handling and distribution
institution-building

Before discussing the characteristics of the individual project classes, it is useful to recognise that the appropriateness of any particular project class for a particular country depends on the internal environment at a specific point of time. The more important features of such an environment are the country's development objectives, be they political or economic and social; the country's stage of development, for example, poor food-deficit countries with *per capita* income levels of, say, less than US$500, versus the more economically advanced developing countries; the extent of development or underdevelopment of resources, such as water, arable land, natural pastures, fishery resources, marketable forest species; and the capacity of the country's institutions.

The more important factors to consider in the selection of alternative project strategies involving particular project classes are listed below.

1 Degree of intensity of use of production factors, capital, land, labour.
2 Investment cost per beneficiary, or unit area.
3 Gestation period of investments.
4 Foreign exchange implications.
5 Ease or difficulty in implementation.
6 Impact on the country's more disadvantaged groups, such as the rural poor and the landless, and the nutritionally vulnerable groups.
7 Impact on the country's recurrent budget after the project becomes operational.
8 Complexity of and time required for feasibility studies.
9 Demands for costly expatriate technical assistance.
10 Contribution to the building up of a local management cadre.
11 Contribution to the development of farmer skills.
12 Opportunities or leverage afforded to influence government policies that impinge negatively on the agricultural sector.

Thus, for any particular country, it should be possible to list project possibilities and screen them, using the above criteria to form a broad initial judgement about the rough order of priority that should be afforded a particular project class. Quite clearly, all the project classes have some merit for applicability in practically all developing countries. However, the country would be concerned with its overall agricultural investment-mix, and what has to be done immediately, and what can be postponed for a later stage.

Technology-oriented projects

The objective of this class of projects is the modernisation of the country's agriculture through increasing the productivity of existing enterprises. The main thrust is to raise physical output per unit of input. Thus increased yields per hectare, or per dollar invested, or per labour unit employed are the objective. This class of projects is particularly appropriate to those countries who are close to the limits of their new agricultural frontiers, and where man/land ratios are quite high.

For most developing countries, this class of projects offers a substantial potential for productivity gains, if the Japanese and Taiwanese experiences are used as indicators of what is attainable on small farming enterprises. Indispensable requisites of this project class are the existence of a technological package adequately tested and suitable for widespread dissemination in the country's major agricultural zones, and a concurrent bundle of measures aimed at increasing the level of farmer skills so that they can implement the new technology.

Generally, this project class comprises projects with small to modest capital investments but entails a considerable increase in the working capital requirements. It is this latter feature which favours their rapid uptake by the larger-landholding classes and militates against their rapid adoption by the smallholder farmers and those who crop rented land.

Although the technology itself may not depend on the scale of operations, mainly farm size, there are two major constraints to overcome if the project's objectives specifically include attempts to reach the smallest farmers and disadvantaged groups. They are first, to ensure that the credit requirements to meet the increased working capital levels are in fact available, and that the farmers are taught the management skills called for by the new technology. Secondly, measures must be taken to ensure that the intended beneficiary does in fact receive the benefits of the surplus production generated, which might otherwise be siphoned off by landlords, and by intermediaries in the marketing chain.

This project class includes the following: projects aimed at the introduction of superior varieties of seeds and planting material; introduction of new and superior livestock breeds for meat and/or milk and new or improved livestock raising systems; introduction of new fishing equipment and techniques to exploit the resources more fully. More commonly encountered are projects designed to raise the level of purchased farm inputs in agriculture, mainly fertilisers, pesticides and rodenticides, and may include input distribution systems. Projects aimed at farm mechanisation fall in this category, and include small farm tractors and cultivation equipment. The harnessing of small water supplies, entailing minor irrigation systems, pumping from streams or wells, is included here because the technology of water management is a major feature of such projects.

A striking example of this class of projects is those concerned with seeds, involving production, distribution and use of high-yielding varieties and purchased inputs. The objective here is to change the technical production potential. The introduction of a disease-resistant strain of a crop, e.g. the dwarf coconut variety in Jamaica, which is resistant to the lethal yellowing disease that kills the tall coconut palms, is such a type of project. Similarly in livestock projects, the introduction of a new legume could well set the basis for a new pasture-improvement programme. This was the core of an early World Bank livestock project in Uruguay. A key determinant of success or failure of this class of project is the willingness of the farming population to undertake risks, as this affects the rate of acceptance of the new technology. Another factor is the state of readiness of the new technology for widespread promotion. Thus, Hirschman concludes from his study of this project and one in Pakistan that the Uruguayan and Pakistani projects suggest:

> a simple hypothesis on the kind of projects whose course is likely to be particularly marked by technological uncertainty and ignorance: both are closely enmeshed with the natural resources of their respective countries, and the voyage of discovery consists in the acquisition of detailed knowledge of these resources and in the needed adaptation of general scientific and engineering knowledge to them.[1]

Broadening the resource base

This class of projects aims at bringing additional physical resources into production, mainly water and land. These include irrigation projects – both surface and underground water schemes; drainage schemes; watershed-type projects, involving erosion control and reforestation; flood control; salinity control; and settlement projects involving new land. This group includes the 'glamour' projects in agriculture. Large irrigation projects are favoured by governments, because they lend themselves to vote-catching ceremonial openings. They are liked by donor countries and international lending agencies because they are clearly visible signs of aid; they lend themselves readily to feasibility studies by consultants, and supervision of construction is relatively simple. They can be, however, powerful development tools since they contribute a higher level of security of production against erratic rainfall patterns.

Large irrigation projects in particular are highly capital-intensive, and their origin is frequently the engineer's desk. The problem arises because the direct output, water, is not a tradeable good. The water becomes an input into the production of a cereal grain or a livestock product. The primary reason for shortfalls in performance is the lack of a multidisciplinary approach to this type of development. Thus the agronomist has not been involved in the question of the cropping programme or crop-water use requirements. The agricultural economist has not been involved in the preparation of farm models or in identifying markets and establishing the basis for price assumptions. The organisation specialist has not been involved in organising farmers at the ground level for acquiring inputs, for water control and in tying together all the parts of the project for implementation.

Even more significant is the fact that the actual delivery of water at the individual farm turnouts, from the primary and secondary canals, falls into a 'no man's land' with no agency or organisation assuming clearcut responsibility. It is the absence of this indispensable complementary investment, including land levelling, that provides the graveyard of many irrigation projects. In fact this situation is so widespread – India, Thailand,[2] Indonesia – that it has given rise to many second-generation irrigation projects, which have as their main objectives, provision of the required investments at the farm level to ensure fuller use of the available water.

These deficiencies aside, this class of project can raise new frontiers of production for many countries. They are clearly justified where the farming population pressure on land is high, e.g. in Indonesia where average farm size in Java is less than 0·5 ha, and rice is the main crop. They are less justified in countries with adequate and reliable rainfall conditions. Generally, one should adopt a cautious approach to this class of projects, and a decision to invest should be taken only after other alternative approaches, involving, for instance, low capital-intensive development of rainfed agriculture, have been fully assessed.

A further characteristic of this class of projects is the socially divisive feature it may introduce. The number of beneficiaries is limited, and the investment per family is usually high. They have long gestation periods requiring up to 12 years or more for the projects' full development benefits to be realised. Long feasibility studies are required, and the hardware involved usually means a high foreign exchange component. The organisation of farm-level services and farmer training to ensure sound water management is a slow process. From a policy standpoint, it does raise the issue of whether, technical alternatives permitting, it is not better to have several small schemes instead of one huge facility, so as to increase the income distribution impact of this type of development.

Improved status of disadvantaged groups

This class of projects is oriented towards deliberate changes in the economic and welfare status of special groups. The dominating consideration is not as in the foregoing case, one primarily of increasing production but rather is a conscious public decision to intervene in the market process and change the ownership structure of the factors of production and to channel the projects' benefits towards designated target groups. This class of projects takes several forms, and is best represented by projects for agricultural credit, land settlement, land reform, nutrition and integrated rural development.

This class of projects has its genesis in the fact that in many socieies, even those in the classic take-off stage like Mexico, growth has not been accompanied by development. Certain areas or groups in the country remain immune to the process of economic growth – best exemplified by the north-east of Thailand and the north-east of Brazil. Both countries experienced tremendous economic growth rates that exerted relatively little or no impact on the regional income disparities.

Possibly, the Malaysian experience is most relevant. Thus Malaysia's new economic policy, embodied in the Second Malaysian Plan (SMP), represents a deliberate government effort to restructure Malaysian society, in order to correct the imbalances in income distribution, employment and ownership and the control of wealth between different segments of the population.

The policy has its roots in the following consideration, namely, that the basic inequality in income levels between rural and urban communities coincides with the polarisation of ethnic groups by economic roles. For example, Malays, some 48 per cent of the total population, account for 83 per cent of employment in agriculture, with the figures for the Chinese being 38 per cent and 11 per cent respectively. The standard of living of Malaysia's rural population, especially of the predominant smallholders' sector, grew only slowly in recent years, largely because of slower growth of productivity in agriculture and stagnation or decline of prices of products, especially rubber.

The new economic policy reflects current thinking that the laws of a free market economy would contribute to a further widening of this income gap. Therefore, conscious and fundamental steps are proposed to correct this situation. Price and subsidy measures are considered inadequate to achieve this end, and structural changes aimed at the very sources of income must be introduced, e.g. redistributing ownership and control of the means of production in the modern sector of the economy.

A consequence of this policy decision was the elaboration of specific investment proposals in agriculture and other economic sectors designed to achieve these objectives.

Agricultural credit projects Though included in this class of projects, agricultural credit projects could sometimes have a predominantly production objective. In many countries, the granting of credit to small farmers has traditionally been tacitly viewed by both the government and recipients as an income-transfer mechanism. The impact on increased production and net farm incomes has been questionable, as a large part of the credit would be used by the recipient for consumption purposes. Iran and Jamaica provide striking examples of the possible dualism of the objective of agricultural credit projects.

Thus, in Jamaica, an agricultural window was opened in 1969 by the Jamaica Development Bank (JDB) – a government bank for industry, tourism and housing – which would provide funds for the development of large commercial farms (over 40 ha) at commercial interest rates. Meanwhile the Agricultural Credit Board, a long established institution, would continue to serve the small farmer sector (farms of less than 40 ha), at concessional rates of interest.

Iran established the Agricultural Development Fund of Iran, with the expressed objective of providing loan funds to the larger enterprises. The smaller enterprises would continue to be serviced by the Agricultural Credit Bank, whose loan repayment record has been far from satisfactory. It is important also to appreciate the reasons for Iran's approach. The land reform programme which Iran had undertaken during the 1950s and 1960s failed to exert a noticeable impact on agricultural production. Iran, therefore, opted for undertaking agricultural production on large-scale enterprises, termed agro-businesses. The Fund, therefore, was created as an instrument to promote the government's agricultural production policies and was later converted into a Bank.

The Caribbean Development Bank in Barbados also carries on its agricultural lending operations along similar lines. The Bank's headquarters staff prepares and appraises loans directly from larger farmers and estates in its member territories, and charges commercial rates of interest. However, the Bank has helped establish agricultural lending institutions within several territories, e.g. Grenada, St Lucia, Dominica, to which it provides low-cost loan funds for on-lending at concessionary interest rates to small farmers.

There appears to be merit in separate institutions catering for large versus small farmers, where a dualistic type of agriculture prevails. The lending problems are sharply different. The main differentiating factor appears to be that in the case of the large farmers, the role of credit is mainly that of a lubricant in an already dynamic context. The small farmer's static situation arises from many factors, and requires a comprehensive package approach in which credit is only one element. Apart from the larger staff needs of institutions dealing with small farmers, there are complementary measures required – introducing new technology, better land-tenure conditions, assured markets, facilities for provision of inputs, etc. – if credit is to have an impact on production by small farmers.

The high administrative costs of credit projects involving small farmers and the risk element arising from crop losses must be resolved if many small farmers are to be reached. Furthermore, institutional credit limited purely to production credit and not consumer credit will always mean that the small farmer must maintain his credit links with the intermediary. In fact, apparently high intermediary interest rates may well reflect the degree of scarcity of capital, and the high risks of intermediaries serving farmers in small localised areas. The standard objective of institutional credit sources of trying to by-pass the intermediaries may well require a re-evaluation. The issue could well be: how to incorporate the intermediary into a formal credit system, thereby forcing him to reduce his on-lending rates to farmers?

Land-settlement projects Whether in Asia, Africa or Latin America, land-settlement projects usually have one common feature – namely, they are designed to settle previously economically disadvantaged farmers or landless farmers on new plots of land. Although production considerations naturally exist, social justice appears to be the main justification. Sometimes the social element is tied in to a political and security objective. In Nepal, for example, the government's land-settlement policy called for the resettlement of landless farmers from the crowded 'hills' to designated areas on a thin strip of land along the Indo-Nepalese border. In Guatemala, a land-settlement project in the north of the country, had as an objective the raising of the population level of the part of the country bordering with Mexico.

Examples abound where countries have undertaken such projects for mainly social reasons, e.g. Malaysia, Kenya, Trinidad and Tobago. The Kenya Government had by 1965 purchased some 0·5 million ha of former European-owned land for the ultimate settlement of approximately 34 000 families. As de Wilde observes:

> The programme was in any event conditioned by political as well as economic considerations. Its basic concern was not with the maximization of output but with the accommodation of more Africans on the land that was previously owned by Europeans. However, the permanent objectives of giving Africans more land and putting more Africans on the land have had to be reconciled as far as possible with the need to maintain and even increase the contributions which the European farms selected for settlement had been making to commercial production and exports.[3]

The Government of Trinidad and Tobago undertook the Crown Lands Project to settle disadvantaged families on government-owned lands, with the expressed purpose of raising the annual income levels of beneficiaries to predetermined levels and thereby reducing the disparity of urban versus rural incomes.

Possibly the largest single constraint facing this type of project is its heavy cost and, therefore, the low prospects for replicating such development. This is true even where land-ownership patterns are such that government does not have to stand the financial burden of land purchase for the settlement. Indonesia provides a classic case where settlement policy has aimed at relieving Java's population pressure and developing the outer islands. Myrdal observed that

> financial limitations are likely to continue to be the most severe obstacle to the programme. Even though the basic services provided at the settlement – land clearance, roads, housing, health facilities, assistance during transitional period, are meager and of poor quality, and much of the work is supposed to be done by the settlers, the cost of colonization is high.[4]

Generally the evidence indicates that the most successful type of colonisation is where settlers immigrate spontaneously, and government provides guidance and the minimum infrastructure requirements, e.g. in Mindanao, in the Philippines and also in Sri Lanka. Apart from the fact that per settler costs can be high (averaging over US$12 000 in Malaysia), the problem is that the classic type of settlement project benefits relatively few families and really does not make a significant impact on alleviating the problem which it is designed to remedy. New settlers must take on a debt burden at levels to which they are unaccustomed, as they must repay both the land and land development costs. When such settlement revolves around the establishment of tree crops, it could require fully a decade before settlers begin to enjoy a sizeable net cash surplus.

In Latin America, the same arguments hold true. Thus Carroll points out that:

> the establishment of an adequate infrastructure in new areas not only requires considerable capital, but there is also a time cost and an element of insecurity which delays the maturing of benefits.[5]

Where *latifundia* exist side by side with *minifundia* as in Latin America, Carroll argues that the land-settlement (colonisation) approach does not alter the basic distribution of property power and income; at best a supplementary measure of land-tenure reform in *latifundia* areas, it is used by a number of countries as a diversionary or delaying measure to avoid massive redistribution reforms.

In most countries agricultural planners face land-tenure problems in project areas. There is no doubt that an unfavourable land-tenure situation is a disincentive to tenants even in the use of current purchased inputs, such as fertilisers, and more so in undertaking longer-term capital investment.

In this situation, there is little that the project analyst can do except to try to convince political leaders of the costs of inaction. Meanwhile international institutions, including financing ones could, to some degree, use loan policies as a leverage to induce action on the part of governments. This is not so easy, because the financing of a land-redistribution programme represents the internal transfer of a resource which would not normally qualify for external financing.

Rural development projects These now constitute the major thrust of lending institutions, such as the World Bank, in the fight against poverty.

The rural poor are estimated to number about 700 million, with the heaviest concentration on the Indian sub-continent, Indonesia and Africa, south of the Sahara, parts of Latin America and the Caribbean. Poverty, however, while being most severe in the rural areas also covers urban areas.

Two main features set these projects apart from other classes of agricultural projects. The first is that there is a conscious effort to give a distributive bias to the projects output by having a broad-based group of beneficiaries as the target group. The second is the multisectorial approach frequently employed in the project's concept and formulation. These are briefly discussed.

Income-distribution considerations would tend to locate the project in an area with masses of low-income farmers. While the direct number of beneficiaries from the project's production components may be modest, the total number of beneficiaries falling within the 'influence zone' of the project and benefiting from it may be considerably larger. Thus in one such project in the Philippines, the beneficiaries were estimated to total more than 200 000 people. In one such project in Upper Volta, some 270 000 people or 30 000 families would benefit from the project works.

Implicit in the multi-sectorial approach is the realisation that, while the rural poor are strongly dependent on agriculture for their livelihood, the determinants of 'quality of life' embrace much more than agricultural pursuits. Thus, project formulation gives considerable weighting to a design that emphasises the inclusion in the package of components and services, normally a public responsibility. The need to tackle the complexity of poverty also requires that intersectorial linkages be explicitly recognised in project formulation.

First, *directly productive* activities would provide the core around which the project should be assembled. This is important, because any self-sustaining growth must be based on increases of production, which would generate savings for reinvestment. Thus the area's physical resources base should be determined by any required surveys. Possible growth poles should be identified and possibilities for raising the existing levels of production in the area by introducing new technology should be fully explored much as in any other type of agriculture project. Such components may include small irrigation facilities and special crop development programmes, including credit and support services.

Second, *production support* activities would be designed to remove constraints to the area's overall economic development. Roads, bridges, port facilities, warehouses and storage facilities are all possibilities. Selection of components should be based on the contribution that they would make to the area's economic growth rather than to immediate but temporary employment generation though this could well determine the method of construction.

Third, *social support* elements appropriate to the area's needs, including items such as health facilities, hospitals, dispensaries, clinics and disease control programmes; educational facilities; village drinking water supplies; and community centres. It is important in this context, to know clearly how existing services are administered; the location and extent of their clientele; the state of repair of existing facilities; the staffing and manning situation; the budgets – capital and operating; arrangements for maintenance of these services, and the extent to which beneficiaries are required to contribute to the cost of such services.

While on the whole, governments tend to be sympathetic towards making resources available for health and disease control programmes (e.g. malaria eradication), they tend to show a lesser degree of commitment to projects of a preventive nature. Village water supplies, falling in this category, therefore warrant special treatment. On economic grounds, increases in labour productivity should arise from improved water supplies. Moreover, the provision of an accessible water supply could, in many societies, make for considerable time savings in the daily chores of fetching water from remote areas. In Nepal, for example, residents in the hill areas, must spend several hours a day in the drier periods of the year, fetching water for domestic purposes.

Because of the increasing importance of rural development, the next chapter examines the approach taken by Mexico in its fight against rural poverty and discusses the experience gained so far in this type of project.

Improved post-harvest handling and distribution

These are essentially marketing and storage projects designed to provide more orderly marketing of products, and thereby reduce seasonal price fluctuations. Alternatively, where the crop is an important source of export income, the government may wish to intervene directly in the marketing operations. Their implementation usually calls for a storage and distribution network. Inventory financing calls for heavy working capital requirements. This type of project results usually from a decision on pricing policy aimed at keeping prices within range for the population, especially the urban part. Sometimes the intervention is geared to protect the interest of producers against the so-called wicked intermediary. Alternatively, a frequently encountered objective is to keep food prices low for the urban consumer.

Marketing projects may also aim at providing infrastructure directly benefiting the farmer, e.g. farm to market roads, drying floors, and on-farm storage. This class of projects generally requires a strong commercially oriented management, which most government civil service administrations cannot supply, and can sometimes impose heavy drains on government revenues. The largely commercial orientation of such projects usually means that they can be more efficiently executed by the private sector with its dominant profit motive.

The potential of this class of projects for the systematic reduction of post-harvest losses has not been explored. In many developing countries such losses are claimed to range from 15 to 30 per cent of harvested production, especially of food grains. Losses from perishables are considerably higher. Project planners and governments should give much more attention to this in their design of a country's food-production strategy.

Institution-building

The project approach to aid for agriculture in developing countries has given considerably more attention to immediate increases in agricultural production than to the build-up of an institutional structure capable of sustaining such increases. While international agencies and aid-donors are conscious of the problem, and the projects financed do frequently have some component, e.g. training, that benefits institution-building, yet the process of project appraisal does not take this aspect of projects seriously into account.

By institution-building we imply deliberate measures designed to strengthen local institutions so that they can undertake successfully development projects and programmes. It is highly questionable whether this can be achieved by piecemeal considerations inherent in the project approach. The whole problem has to be seen in a different context. The proper focus is the overall development constraint arising from the absence of strong national institutions in the countries. The problem is particularly serious in the United Nations group of countries termed the 'most seriously affected countries'. Measures to increase the 'absorptive capacity' of such countries to undertake agricultural investment projects require urgent attention but development institutions have not yet found plausible solutions to this.

In fact, in many such countries, the available aid-flows far exceed the countries' capacity to utilise such aid effectively. This results in project loan or credit funds remaining undisbursed, and/or project targets not being realised.

Any concerted plan attacking the institutional bottlenecks in these countries requires a three-tier approach. The first is at the country planning level. The critical need here is to develop capacity for project generation. Since this activity cuts across many government departments and agencies, there is need for strong mechanisms for coordination. The tasks at this level should, in fact, go beyond project generation to include provision of supporting services to the management of approved projects ready for implementation. Such factors as recruitment of consultants, letting of bids for contracts and obtaining release of foreign exchange, are subject to bureaucratic snarls, resulting in delays in project implementation.

The second level of intervention would be at the project-management level. The common approach to this problem of shortage of local managerial skills has been to provide massive injections of expatriate technical assistance, primarily from developed Western countries, through the construction period of the project, with the expectation that local nationals would receive on-the-job training. There is room for new approaches both by the developing countries themselves and the development institutions and aid-donors. The relevant body of experience that is meaningful for agricultural project-development work lies in the developing countries themselves. The development agencies, working with the countries, could categorise this experience in relevant fields. A few examples are water management, livestock, forest industries, agricultural credit, land settlement, rural development and seed projects. Selected countries could undertake training programmes for potential project managers from other developing countries. Financing of such training could be done by aid-donors, by regional development institutions, and by the countries themselves. Such an approach could well set the basis for a more formalised and systematic exchange between countries of their experience in agricultural development projects.

The third level of intervention would concentrate on farmers. There are two aspects to this; one is support for farm-level organisations, where again shortage of managerial skills is always acute; the other is the improvement of farmer skills so as to be able to utilise fully the new technology.

While conceding that greater emphasis on institution-building could, in the short-run, reduce the initial flow of a project's benefits, the medium and longer-term considerations of maintainable increases of production should not be neglected.

References

1 Albert O. Hirschman, *Development Projects Observed*, The Brookings Institute, Washington, D.C., 1967, p. 41.

2 *Note:* in an effort to overcome this serious problem, the Government of Thailand transferred the Royal Irrigation Department from the Ministry of Natural Resources to the Ministry of Agriculture, so as to associate irrigation more closely with agriculture.

3 John C. de Wilde, *et al.*, *Agricultural Development in Tropical Africa,* Vol. II, *The Case Studies*, Johns Hopkins University Press, Baltimore, 1967, pp. 188–9.

4 Gunnar Myrdal, *Asian Drama – An Inquiry into the Poverty of Nations*, a Twentieth Century Fund Study, Vol. III, Penguin Books, 1968, p. 9.

5 Thomas F. Carroll, *Issues of Financing Agrarian Reform: The Latin American Experience*, World Land Reform Conference, FAO, Rome, WRL/66/5, 1966, pp. 18–19.

4 Rural development – the Mexican approach and discussion

The success of rural development programmes and projects in a country requires first and foremost, a strong political commitment on the part of the governmental authorities. This arises not only because substantially larger resource-flows must be diverted to the rural areas, but also because the targeted beneficiaries must be selected from among the broad masses of the rural poor. There must be an explicit will to confront squarely the question of rural poverty.

The formulation of a specific rural-development programme, comprising a package of discrete projects, requires the following:

1 Organisational arrangements to ensure full participation of the many development agencies involved in project execution, including decentralised decision-making.
2 The adoption of explicit criteria for the choice of benefiting micro-regions.
3 Guidelines relating to the investment-mix of projects.
4 Arrangements for involving the beneficiaries in the planning process.

The summary description of such a rural development project in Mexico is particularly interesting because it represents the government's clear response to challenges facing the country in the distribution of the fruits of its economic growth.

While Mexico's economic growth rate in real terms averaged about 7·0 per cent between the period of 1960 to 1975 – a performance far exceeding that of most developing countries – the benefits of such growth have not accrued to the poorer strata of the population. Thus, in mid-1973, some 2 million families – over 50 per cent of the rural population and 20 per cent of the total population – had average incomes of less than US$800 per family in contrast to the average family income estimated nationally at US$2 700; government estimates assume that a per family income of about US$960 is required to provide minimum acceptable living standards for families in rural areas.

Mexico's rural population of about 20 million comprises three main groups, private farmers, landless families and *ejidatarios*. By law, the *ejido* is essentially a group of families with joint and inalienable tenure rights to land.

While previous government policies had, during the period 1935–41, pursued an extensive land-distribution programme, thereby favouring *ejidatarios*, later policies pursued a different strategy favouring industrialisation and the development of large-scale irrigation projects. This latter approach had positive benefits on agricultural production which showed a growth rate averaging 5 per cent over the period 1950 to 1970. However, this growth occurred at the cost of widening disparities between the more privileged and the poorer strata of the rural population.

In 1970, government introduced major policy changes in its approaches towards rural development by substantially increasing resource flows and purposefully channelling them towards the poorer rural areas of the country. A three-prong approach comprised, first, a new focus on poorer groups in on-going river-basin development areas, with the provision of low-cost productive and social investments; second, the strengthening of capacity of different agencies to undertake rural-development activities through the creation of separate rural-development departments in these agencies; and third, the introduction of a new Programme for Integrated Rural Development (PIDER).

The PIDER programme initiated in 1972, has the primary objectives of developing micro-regions suffering from acute rural poverty by providing investments and services in selected areas, so as to raise rural living standards, to increase employment opportunities and improve basic social infrastructure. The average micro-region has three municipalities with about 50 000 people.

The main distinguishing feature of the PIDER programme is its commitment to increasing the productive capacity and income redistribution in favour of the most economically disadvantaged groups. PIDER plays a strong role in coordinating the development activities of participating agencies[1] in the selected micro-regions. The Ministry of the Presidency which exerts final authority on the investment budgets of all government agencies, has the main responsibility for the execution of the PIDER programme. This arrangement ensures coordination of the activities of various agencies in the micro-regions.

The three-tier organisation of PIDER is shown in Figure 4.1. At the federal level the Central Coordinating Committee, comprising senior officials from participating agencies and headed by the Presidency's Director of Public Investment, is the highest policy-making body of PIDER. This body approves programmes of micro-regions and addresses itself to problems of inter-agency coordination. At this level, a Permanent Working Group, made up of representatives from various agencies, provides technical support to the Coordinating Committee.

At the state level, the Governor chairs a state committee, comprising state offices of federal agencies, which coordinates the execution of the programme of the participating agencies in the micro-regions. The state committee also has its technical group who directly supports the PIDER direction for the state.

At the village and *ejido* levels, there is an involvement of the municipal chief and *ejido* and village authorities, whose task is to organise the communities for local participation in the PIDER investment programmes.

Criteria for selection of micro-regions

Individual micro-regions in PIDER's programme for investment were selected on the basis of the following criteria: population density, productive potential, poverty levels and lack of basic infrastructure such as roads, markets and drinking water. In delineating micro-regions,

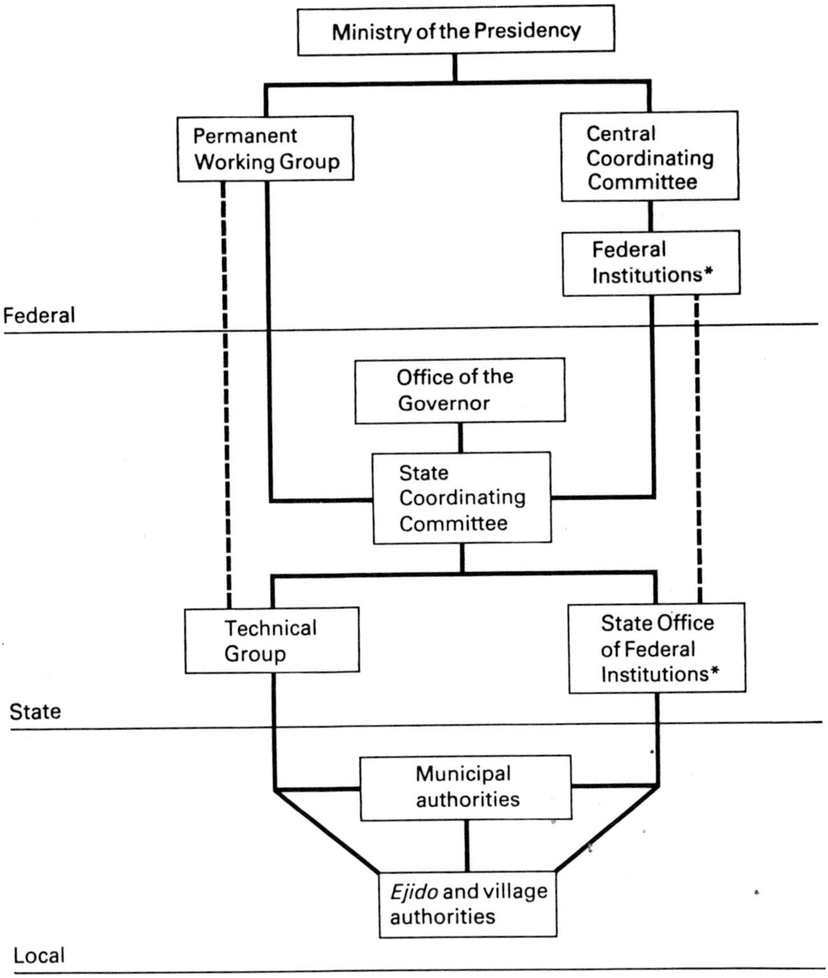

Figure 4.1
Mexico: Rural development project – Organisational chart, federal, state and local levels.

*Federal institutions represented at the federal and state level include: Ministry of Agriculture, Ministry of Finance, Ministry of Public Works, Ministry of Water Resources, Ministry of Health and Assistance, Department of Agrarian Affairs and Colonisation, Federal Electric Commission, National Corporation for Basic Marketing, Committee for Administration of the Federal School Construction Programme, National Arid Zones Commission, National Indigenous Institute, National Institute for Rural Community Development and Low Cost Housing, the *Ejido* Bank, and The Guarantee Loan Fund.

PIDER employed the following yardsticks:

1 total population of about 100 000;
2 total area of about 7 000 km;[2]
3 a maximum of 75 localities between 300 and 3 000 persons;
4 existence of one or more growth poles;

5 a high proportion of the economically active population with an income of less than US$1 000 per annum;
6 a high index of under -employment;
7 adequate development potential; and
8 inadequate social infrastructure.

During the appraisal of the World Bank financed project, described below, refinements were introduced in the criteria for selection of micro-regions. Selected ones had to meet certain minimum threshold criteria involving population density. The poverty criterion, in terms of income, was set at 40 per cent of the economically active population with incomes less than US$500 per year. For under-employment, the criterion was set at two per cent of the labour force economically active less than six months per year. Productive potential was assessed on the basis that the investment programme should yield a minimum of ten per cent economic rate of return on directly productive and productive support investments.

Investment programming at the village/*ejido* level

Investment programming at the village/*ejido* level is undertaken by visiting teams of state and federal experts. Detailed baseline data, assembled prior to the visit, are evaluated by the programming team. The range of data includes population and income data, cropping patterns, availability of existing resources, e.g. groundwater, stream flows, soil and rainfall patterns and basic infrastructure and services, such as schools, clinics, etc. Discussions are held by the team with villagers who propose their own development priorities. These are weighed against the realities of the data base, and PIDER's own development strategy for the micro-region. A multi-sectorial village-level investment programme is then developed and these are aggregated for the different villages into micro-region development programmes.

Table 4.1 provides a summary description of the main coordinates of the investment package for 15 micro-regions covering the period 1973–6. The data gives the range and average for each coordinate.

The ecological zones represented in the table ranged from arid to wet tropics, including temperate and dry tropical conditions.

TABLE 4.1 Mexico: Rural development project – Coordinates of investment package

Item	Unit	Range	Average
Population	'000	24·7– 102·6	50·9
Area km^2	'000	0·6– 17·4	8·5
Investment distribution			
directly productive and productive support	per cent	72·7– 90·9	82·0
social infrastructure	per cent	8·5– 25·2	18·0
Investment *per capita*	US$	113.0– 658.0	197.0
Before development family income	US$	230.0– 575.0	420.0
After development family income	US$	550.0–1 935.0	1 160.0
Financial rate of return	per cent	15·5– 33·9	22·5
Economic rate of return	per cent	10·8– 23·2	16·0

One feature worthy of note in the data is the very modest level of average investment *per capita* – about US$197·00. This is inevitable if limited resources are to reach masses of eligible beneficiaries.

The magnitude of the programme is seen from the following: by the end of 1974, PIDER had invested US$133 million in the initial micro-regions – of which 45 were by then in operation. Investment requirements programmed for the completion of these 45 regions totalled US$450 million over the period 1975–7. Government plans called for the development of a further 55 micro-regions requiring additional investment of US$650 million. The development costs of the 100 micro-regions would thus require total investment of about US$1 200 million. The programme at full development would aim to benefit some 5 million poor rural people, or about 25 per cent of the rural population.

Here again, the average investment per beneficiary in this global programme of US$240 appears reasonable and lies fairly close to the US$197 above for 15 micro-regions, and the difference must be explained by overhead costs, including administration. Yet, even at such modest investment levels per beneficiary, the total investment resources, in cash, required to cover the whole population would be about US$4 800 million. This calls for massive governmental resource flows into the rural sector through the PIDER channels, particularly since only about one-half of the programmed investment is deemed recuperable. The question of cost recovery is briefly discussed later.

The profile of a micro-region

The El Llano micro-region is located in the State of Aguascalientes – one of the smallest states in Mexico. Of the 388 000 total state population, about 200 000 live in the State's capital city. The El Llano micro-region comprises four municipalities with a total of 75 000 rural people, who are over 75 per cent dependent on agriculture and related activities. About three-quarters of the working population earns less than US$500 per year, and thus from a poverty standpoint meet the PIDER's selection criteria.

The region's resource base, though favourable, is under-utilised. While some semi-desert areas prevail, a large portion of the region offers considerable potential for increasing both cropped area and yields. About 8 per cent of the total land-area enjoys irrigation, both surface and underground. The livestock resources – 170 000 head of bovine cattle, 40 per cent of which are dairy animals – are under-utilised due to various constraints such as serious overgrazing, inadequate supplementary feeding and inadequate technical services.

About 80 per cent of the labour force are mainly subsistence farmers. Farm employment activities account for an annual average of 120 days per farmer, literacy rates are high and the road network is described as ample.

The PIDER objectives and strategy for the micro-region are to increase permanent employment in agriculture and industry, and to raise labour productivity and income levels. The development strategy recognised the differing requirements of the main sub-regions.

In sub-region 1 – the Pabellón irrigation district – the plan aims at yield-increasing technology and water-saving practices, and at inducing shifts in the cropping pattern to labour-intensive, high-value crops, e.g. peaches, grapes, pears, peppers, potatoes and alfalfa production for expanding the dairy industry.

In the other sub-regions, the plan aims to raise yields of crop and livestock under rainfed conditions. The investment package would provide for further water exploration and well-drilling; soil- and water-conservation works including small earth dams for rainwater storage for cattle and irrigation, and social infrastructure, e.g. drinking-water supplies and electricity.

The PIDER plan to implement the strategy called for an investment of US$6·1 million over the period 1973–7. The breakdown of the proposed investment involved the following.

Of the total investment:

1 55 per cent would be allocated for directly productive infrastructure where irrigation and livestock-related projects accounted for 85 per cent of this amount. Water conservation works and rural industries would account for the rest.
2 22 per cent for productive support items, especially agricultural extension and demonstration.
3 21 per cent of social infrastructure divided evenly between electrification, drinking water and materials for self-help projects.
4 2 per cent for programme coordination.

The investment programme would directly benefit 10 700 families in 100 communities. Incremental employment would include: 1 400 man-years of temporary employment over four years, and 500 man-years of permanent employment.

The findings of PIDER analysis for the *ejido* San Antonio – comprising 82 *ejidatarios* and 1 000 inhabitants – show the following expected income changes, by category of activity. The income per family (of seven members) would rise from US$185 before development to US$350 after development under rainfed agriculture (maize and beans). For irrigated agriculture (maize, beans, chillies, alfalfa), the corresponding figures would be US$146 and US$500 respectively. Increments in beekeeping and cattle operations would be US$150. Income per family from all operations would triple from US$340 to about US$1 000.

The project

The project submitted to the World Bank for financing covered the investment implementation requirements of 30 micro-regions. The project had a total cost of about US$295 million, and a World Bank loan was approved for US$110 million in 1975. The Inter-American Development Bank also financed a further 15 micro-regions on a parallel-financing basis.

About two-thirds of the total project costs (66 per cent) were allocated to directly productive investments; productive support investments received 22 per cent, and social infrastructure 12 per cent. Provision was made for project evaluation. These are briefly discussed.

The directly productive investments comprised irrigation, livestock development, soil and water conservation, fruit production, and other investments, e.g. beekeeping, fishery and forestry development, and a large component of farm development credit.

The irrigation component (17 per cent of total project costs), aimed at the development of small-scale irrigation of 30 000 ha at an average cost of US$1 200 per ha. Each proposed irrigation sub-project would be subject to technical and socio-economic feasibility studies in accordance with approved guidelines. Conditions were set upon the sub-loans so as to ensure control over both the unit costs and the choice of beneficiaries. Thus, for projects up to 25 ha, the cost per family was limited to US$4 000. For smaller works, the total cost of a single project would not exceed US$32 000 and benefit a minimum of eight families. In larger projects (up to 2 500 ha) families with a pre-project net income of more than US$2 500 would not be eligible.

Livestock development (10 per cent) would be undertaken on 800 *ejidos*. Beef-cattle development with emphasis on low-cost livestock facilities would involve land clearing, pasture establishment, fencing and watering and handling facilities.

Soil and water conservation (4·9 per cent) would involve construction of soil- and water-conservation works using labour-intensive methods. Detailed conservation plans on a watershed basis were required for each village or *ejido* area. The construction of contour ridges, terraces and checkdams would be undertaken over 91 000 ha.

Fruit production (4·9 per cent) would be supported on 9 500 ha and involve orchard infrastructure and initial planting of fruit trees.

Other productive investments (1 per cent) would involve mainly reforestation activities.

Farm development credit (29·6 per cent) would provide largely for medium- and long-term credit needs associated with irrigation investments and livestock purchases. Mexican credit institutions had recently developed criteria and improved their capabilities for lending to low-income farmers.

The productive support activities, financed under the project, would comprise feeder roads, rural marketing, farmer organisation and training, extension and field demonstration, and rural electrification.

Feeder roads (6 per cent) about 1 900 km of road construction would be supported under the project, with emphasis on low-cost feeder roads and labour-intensive methods of construction.

Rural marketing (0·7 per cent) would involve the construction of 60 small stores and five warehouses for the purchase and storage of farm produce.

Farmer organisation and training (2·8 per cent) would support expansion of staff within the micro-regions.

Extension and field demonstration (7·1 per cent) would support field demonstrations of the recommended technical packages and improve the effectiveness of the extension services operating in the micro-regions. Technical packages, adapted to low-rainfall conditions (less than 750 mm) are not as well developed as those for the higher rainfall and irrigated areas. The aim was to disseminate the use of low-risk packages among farmers in the involved micro-regions.

TABLE 4.2 Mexico: Rural development project – Estimated project cost

	Local	Foreign (Mex$ million)	Total	Local	Foreign (US$ million)	Total	Foreign Exchange (per cent)	Unskilled labour (per cent)	Percentage of baseline cost
Directly productive									
Infrastructure									
Livestock	210	52	262	16·8	4·2	21·0	20	20	10·0
Irrigation	330	110	440	26·4	8·8	35·2	25	30	17·0
Fruit production	65	7	72	5·1	0·6	5·7	10	35	2·7
Soil and water conservation	112	18	130	9·0	1·4	10·4	14	75	4·9
Beekeeping	13	0	13	1·1	0·0	1·1	0	14	0·5
Forestry	20	2	22	1·6	0·2	1·8	10	35	0·8
Fishery	2	0	2	0·1	0·0	0·1	20	20	0·1
Sub-total	751	190	941	60·1	15·2	75·3			36·0
Credit									
Development credit	455	114	569	36·4	9·1	45·5	20	15	21·1
Incremental production credit	198	27	225	15·8	2·2	18·0	12	20	8·5
Sub-total	653	141	794	52·2	11·3	63·5			29·6
Productive support									
Feeder roads	151	11	162	12·0	0·9	12·9	7	70	6·0
Extension and field demonstration	178	11	189	14·2	0·9	15·1	6	0	7·1
Farmer organisation	73	3	76	5·9	0·2	6·1	4	0	2·8
Marketing	16	1	17	1·3	0·1	1·4	5	10	0·7
Electrification	114	38	152	9·1	3·0	12·1	25	21	5·7
Sub-total	532	64	596	42·5	5·1	47·6			22·3

Social infrastructure									
Health	35	6	41	2·8	0·5	3·3	15	30	1·6
Education	61	11	72	5·0	0·8	5·8	15	27	2·7
Drinking water	126	31	157	10·0	2·5	12·5	20	40	5·9
Self-help projects	40	2	42	3·2	0·2	3·4	5	0	1·6
Sub-total	262	50	312	21·0	4·0	25·0			11·8
Evaluation	8	1	9	0·6	0·1	0·7	10	0	0·3
Total baseline cost	2 207	446	2 652	176·4	35·7	212·1	(17)	(28)	100·0
Physical contingencies	158	32	190	12·7	2·5	15·2			7·2
Expected price increases	697	141	840	55·9	11·3	67·2			31·6
Total project cost	3 062	620	3 682	245·0	49·5	294·5	(17)	(28)*	138·8

* Total direct labour represents 34 per cent of project costs, 28 per cent unskilled, 6 per cent skilled and professional labour.

Rural electrification (5·7 per cent) would provide electricity for about 170 000 rural people in 300 of the poorest villages. About half of this would be associated with directly productive investments. An average cost of US$70 per beneficiary was estimated.

Social infrastructure would comprise educational facilities at the village/*ejido* level, rural health, rural water supplies, and materials for self-help projects.
Educational facilities (2·7 per cent) would renovate and construct some 1 150 primary level classrooms, involving 60 000 students.
Rural health (1·6 per cent) would renovate, construct and equip 220 rural health posts and 70 rural health centres.
Rural water supplies (5·9 per cent) would provide rural water systems for about a quarter million people at an average cost of US$50 per person.
Self-help projects (1·6 per cent) would finance the purchases of materials for developing projects on a group basis, e.g. workshops, community centres, etc. The beneficiaries would provide the labour.

Project costs estimates are shown in Table 4.2. Of the total costs of US$294·5 million, the foreign exchange component totalled US$49·5 million or 17 per cent. The World Bank loan thus covered the foreign exchange cost and about 21 per cent of the local cost.

Project's benefits

Since project design and selection criteria paid appropriate attention to the potential of the various ecological zones, the projection of yields must reflect these regions. They interact with the project's main actions, such as irrigation development, farmer training and extension, provision of credit, and input delivery systems. The yield increments projected are shown in Table 4.3 for the major crops, maize and beans, which are the staple dietary foods of most Mexican families.

TABLE 4.3 Mexico: Rural development project – Projected increases in production and marketing

	Base-year production	Incremental production (1 000 tonnes)	Incremental marketed surplus
Maize			
irrigated		60	40
rainfed	1 100	140	70
Beans	70	40	20
Beef	90	25	20
Fruits and vegetables	40	10	5

It should be noted that the large differences in assumptions about yield increments, reflect the inadequacy of technical packages suitable for low-rainfall areas.

The incremental production generated by the project was assessed on the basis of projected yield increments, and the areas devoted to the various types of enterprises. The incremental marketed surplus shown in Table 4.4 assumes that at a low-income levels, a very large part of the additional output ends up being directly consumed by the farm household.

TABLE 4.4 Mexico: Rural development project – Projected crop yield increases

	Unit	Before development	At full development
Maize (irrigated in semi-arid areas)	kg/ha	750	2 500
Beans (irrigated and interplanted with maize in semi-arid areas)	kg/ha	200	500
Maize (1 200 mm rainfall)	kg/ha	1 200	1 700
Beans (1 200 mm rainfall and interplanted)	kg/ha	300	400
Maize (800 mm rainfall)	kg/ha	850	1 100

The marketing of these products should present no problems as the incremental output of maize and beans would represent only about 2 and 4 per cent respectively of Mexico's production of these crops.

The project's national production increases provide no guidance as to the readiness of farmers to participate in the project. This is a function of what producers expect to receive as a result of the project actions – that is, the adequacy of the financial benefits, which is the best practical guide to assessing the project's incentives. These producer benefits are briefly summarised below.

The project would raise producer incomes over an eight-year period from an average of US$440 per farm to US$700 per farm in rainfed areas, and to US$1 100 for families with access to new irrigation works. For a typical *ejido* farmer, with a small, 2-ha irrigated area, the income increment would be as shown in Table 4.5.

TABLE 4.5 Mexico: Rural development project – Impact on producer incomes

Category	Before development	After development
Irrigation (maize)		420
Livestock		
cattle	40	60
bees	10	20
other	10	30
Rainfed		
maize	330	470
beans	40	70
Fruit production		20
Rural industries and other	10	20
Total (US$)	440	1 100

The beneficiaries under this project would total about 1·5 million rural people. The target group would consist of families with incomes ranging from US$250 to US$750, placing them in the lower 50 per cent of incomes in the rural sector and lowest 20 per cent of the total population. The real income of about 120 000 farmers (three-quarter million people) would about double over an eight-year period, while those benefiting from irrigation would triple. A further 350 000 people would benefit

TABLE 4.6 Mexico: Rural development project – Estimated recovery of investment costs

Project component	Total investment (US$ million)	Recovery period (years)	Interest rate (per cent)	Amount recovered (US$ million)	Proportion recovered (per cent)
Directly productive					
Infrastructure					
Livestock infrastructure	21·0	25		10·0	50
Irrigation	35·2	25		17·0	50
Fruit production	5·7	12		2·8	50
Soil and water conservation	10·4				0
Beekeeping	1·1	10		1·1	100
Forestry and other	1·9				0
Sub-total	75·3			30·9	(40)
Credit					
Development credit	45·5	3–12	7·6	45·5	100
Seasonal production credit	18·0	1	12·0	18·0	100
Sub-total	63·5			63·5	(100)
Productive support					
Feeder roads	12·9				0
Extension	15·1				0
Farmer organisation	6·1				0
Marketing	1·4	15		1·4	100
Electrification	12.1	25		12·1	100
	47·6			13·5	(28)

Other infrastructure			
Health	3·3		0
Education	5·8		0
Drinking water	12·5	0	0
Self-help projects	3·4		0
	25·0	0	(0)
Evaluation	0·9		0
Total baseline cost	212·1	107·9	
Contingencies	82·4	42·0	
Total project cost	294·5	149·9	(51)

from improved rural infrastructure. Employment generated would include 84 000 man-years of temporary labour and 30 000 man-years of permanent labour. Further benefits would derive from the social infrastructure, such as better literacy and health standards. The project would further promote the development of self-help activities.

Cost recovery and budgetary implications are especially important in a project of this scale and type. There are two conflicting considerations here, the first is that the larger the share of cost recovery of the investment, the more prospects exist for replicating the project in other project areas with other beneficiaries. The second consideration is the project's basic objective – which is the eradication of poverty and improved living standards. The government's policy on cost recovery takes account not only of budgetary criteria but also the extent to which private benefits are identifiable, and the repayment capacity of beneficiaries. While government would aim at full recovery of the most directly productive investment, the repayment rates would depend on the income levels of beneficiaries. On this basis, an estimated 40 per cent of the directly productive infrastructure would be recovered.

In livestock infrastructure for example, about 30 per cent of the expected beneficiaries with pre-project family incomes of less than US$375 per annum would not be able to repay the investment. Another 40 per cent of beneficiaries with incomes between US$375 and US$475 could repay an estimated 50 per cent. Soil and water conservation are investments of national interest and are not recovered. Production-support investments such as for feeder roads, extension and field demonstration and farmer organisation would not be recovered from beneficiaries. Health, water supply, primary education investments are not recovered. For drinking water supplies, beneficiaries would be required to contribute at least 15 per cent of the construction cost of the system, either in labour, materials or cash. For electrification, beneficiaries would, on installation, pay a 20 per cent cash contribution of the investment cost of the system and all costs of internal household wiring. The other 80 per cent of the system's cost would be recovered through electricity charges.

Table 4.6 shows the estimated recovery of investment costs by project sub-categories.

The project's annual operating and maintenance costs would total about US$13·1 million. It is planned that the contribution by beneficiaries would be about 50 per cent for the overall project, but they would bear the entire operating and management costs of the directly productive activities.

Appraisal Techniques

The appraisal methods used in this project warrant some attention. Whereas the technical, economic and the target beneficiary aspects of a project of this scale, and the diversity of the ecological zones deserve full attention, the single most important consideration in reducing the risks inherent in the project, is the capacity and experience of the PIDER and associated institutions (a total of 14 federal and state agencies) to implement such a complicated project. It was necessary, therefore, to appraise the sub-sectors concerned – a total of 14 including, for example, irrigation, feeder roads, marketing, health, rural electrification, etc. – against the background of the agency normally responsible for undertaking the particular activity. For example, in dealing with rural electrification, the agency is described, the general policies affecting rural electrification are discussed; and the project actions clearly brought out.

For appraisal purposes, the country was divided into six ecological zones – arid, semi-arid, transitional, temperate, dry and wet tropics. Field appraisal was conducted of micro-regions that were broadly representative of conditions prevailing in these zones. For each micro-region, approval for disbursement of loan funds would be based on an investment programme balance, an assessment of the micro-region, organisation and staffing and a threshold economic rate of return of 10 per cent. For 15 of the micro-regions not yet covered at the time of field appraisal, a condition for eligibility was that investment packages with less than 70 per cent productive investments including credit would not be eligible for loan funds.

During appraisal, a more systematic methodology of investment programming at the village/*ejido* level was developed. The objective of the methodology was to make more explicit the income and employment generating aspects of the *ejido* development programme and to test the internal consistency of the investment package. The investments programmed in the representative *ejido* were systematically evaluated using an economic rate-of-return test (above 10 per cent), and estimates that temporary labour required for development works related to the development of that *ejido* could be met from the off-season under-employed labour force in that *ejido*. This approach provides an analytical tool whereby the investment programme, consolidated for the micro-region as a whole, meets adequate financial and labour availability tests. The methodology requires the calculation of financial and economic rates of return.

Experience in rural development projects

The findings of Uma Lele[2] on the African experience in rural development are of particular interest. Although the study embraces several projects in various countries (Cameroon, Ethiopia, Kenya, Malawi, Mali, Nigeria and Tanzania), which were initiated mainly in the 1960s, before a body of thinking had begun to emerge on the whole rural-development process, the validity of the analysis remains intact since the objectives and project actions did not differ substantially from those projects more recently prepared except to the extent that currently formulated projects tend to have a larger component of social services, and to be more deliberately oriented to the poverty target groups.

In the Lele study, rural development is defined as improving living standards of the mass of low-income population residing in rural areas and making the process of their development self-sustaining. Based on this definition, Lele concludes that while the programmes represented a considerable step towards realising the objectives of rural development, they had been less than fully effective in making the process of development of the low-income sector self-sustaining.

Among the more important reasons cited for this are: many projects when formulated had limited objectives, e.g. increasing export-crop production among smallholders. The project planning did not take enough account of broad sector and policy questions, e.g. national pricing policies and their impact on project performance. Programmes were often based on insufficient knowledge of technological possibilities, and their suitability to small-farm conditions. They suffered from poor knowledge of the socio-cultural and institutional environment in which they were to be implemented. The programmes experienced extreme scarcity of trained local manpower.

In evaluating the impact of national land policies on rural development programmes, Lele points out that where acute inequity in distribution of land-rights exists, then a more permanent change in the distribution of land-rights should be an integral part of the rural-development strategy. With regard to technology, food crops suffered most from the research gap – a factor which not only affects rural welfare, but impedes rural growth. Little attention was given to private trading systems since they were viewed as being exploitative and antisocial.

Trained manpower proved a severe constraint on the expansion of the rural services, while a shortage of extension staff presented a major problem. Investment in training in manpower has not matched the investment in rural development programmes and the manpower constraint results in a tendency to concentrate on progressive or innovative farmers. Programme administrators tend to take a paternalistic viewpoint towards development of strong viable local organisations for helping in the delivery of services.

In the African context, the low-income population spans the entire rural sector. If these masses are to be reached, then development strategy must focus on setting a few clear priorities, e.g. raising productivity, generating increased resources, and increasing the indigenous institutional capacity to undertake development. When the peasantry is politically conscious, as in Kenya, a few regional programmes can generate substantial political tensions.

On the institutional and managerial side, these programmes are usually administered by autonomous integrated bodies or project authorities, alongside an existing administrative structure, with the former having a high proportion of expatriate managers. A major problem arises when the project works are completed, and the project implementation must revert to the normal administrative structure. Thus Lele argues the case for improving the performance of the indigenous administrative systems.

Issues in rural development projects – a discussion

The first issue that arises is the degree of priority that should be accorded rural-development projects in any particular country in contrast to other appealing objectives such as rapid increases in food production and/or expanding agricultural exports. It is partly because of this conflict that some governments tend to be rather lukewarm to the undertaking of rural development projects, let alone the fact that many countries are loath to borrow money on commercial terms for this type of project. In fact, many governments prefer to borrow externally for, say, irrigation development instead of rural development. One reason is that the directly productive investments in the former may average about 80 per cent of total project costs as opposed to the latter with, say, 50 per cent.

The irony of the situation appears to be that those countries that are most in need of rural development because of the widespread and endemic nature of poverty, are precisely those who are the least likely to be able to afford the costs of rural-development projects. Costs here refer not only to capital investment costs of the project but also to the recurrent operating costs especially for social services components, for which no satisfactory system exists for cost recovery. Usually, these governments face severe budgetary problems and demands for local cost financing may exceed their reach.

The problem arises mainly in the poorest countries in the world, where threat of famine, or acute food shortage, undernutrition and malnutrition are ever present. These same countries are the ones with severe foreign exchange and debt service problems, which can hardly afford food import purchases in times of severe national stress, be they caused by droughts, floods, late monsoon rains, or other natural disasters. Governments, faced with such situations, are reluctant to commit themselves strongly to rural development involving considerable investment in non-directly productive elements, because of the difficult policy issues involved.

They might feel, for example, that rural development is technically difficult and beyond their administrative capacity; or they may feel that the economic cost in terms of slower growth in production and exports may be too high for the country to bear; alternatively, they may feel that the prospects for success of rural development projects are slight under existing policies and institutional limitations. Politically the government may lack both the will and capability to institute needed changes, e.g. in land reform. Pricing policies long established in favour of the urban dweller, may induce social tensions if efforts are made to change this.

In fact, the very nature of rural development implies changes in the country's power structure, whereby the existing elite power groups are required to reduce their hold on the main factors of production.

While pursuing the concept of rural development as a long-range objective, a first approach may have to lay the major thrust on broad-based productivity gains with the major emphasis on the food-producing belts of the country, using the least capital-intensive approaches possible. In such selected key impact areas, existing administrative structures would be strengthened by deliberate involvement in the development process. A planned rural infrastructure programme would reflect the priorities required to support the production thrust. The national institutions concerned with social services, whose own locational selections for priority attention may not in fact coincide with the production impact areas, would then plan to provide for the impact areas, a minimum level of social services consistent with their national budgetary allocations for such services.

A second issue, partly related to the foregoing, is whether it is politically acceptable or wise for a government to concentrate resources in one selected area, at a level far above the national average, when poverty itself is widespread and not just characteristic of one or two regions. One Caribbean government, for example, would not accept the idea of a rural-development project whose location could have resulted in marked benefits to only a favoured section of the country. In a compromise solution, the project had to be adjusted to include components geared towards providing benefits for other areas of the country.

An area concentration approach can tend to generate divisiveness at three levels – between political representatives of the various regions; between the national administrative units at the centre who may disagree with the allocation of their budgets implicit in the concentrated area approach; and lastly between the populations of the different regions. This situation is especially prone to occur where the effort in one region is of such a scale that it just cannot, for lack of resources, be repeated in other equally deserving regions.

A third issue relates to the degree of emphasis to be attached to the agriculture sector, to the virtual exclusion of other economic activities, in planning for the production core of the project. Project planners should pay considerably more attention to the region's or project area's 'economic base'. The implicit assumption in projects prepared to date is that the economy of rural areas is synonymous with agriculture. This type of partial analysis can only lead to regional solutions that may be at variance with the realities of the situation.

Much more needs to be known about the region's overall economy, its flow of trade, goods and services not only within the project's impact area, but with the rest of the country, and the characteristics of non-farm employment. In the calculation of a project's benefits, for example, the effects of the non-farm income are hardly ever taken account of; yet, non-farm income can be a substantial part of the total income of the small farmer. Even less is known about rural landless workers and the non-farming rural population.

In particular, the role of small-scale industries warrants attention. The findings of a seminar on small-scale industry in Latin America, for example, show that small-scale industries have a greater locational flexibility than large enterprises, since they can make use of local materials and cater to small markets. This could be part of an income strategy aimed at the smallest farmers with non-viable units and the landless so that they can tap the surpluses of food generated by the project's investments.

A fourth issue relates to involvement of the targeted beneficiaries in the project-planning process. It is not uncommon to find that centrally designed plans are imposed on local authorities and that there is a top-to-bottom approach, instead of aggregation of plans at the local level for sanction by the top levels. Fortunately, we are witnessing favourable developments in this regard. In one World Bank and IFAD financed project in Tanzania, specific provision was made for one component called the Village Self-Help Programme, whereby at the village level, credit would be provided for those minor developments proposed by the villagers themselves. The FODERUMA programme of Ecuador, is an excellent example of governmental effort to involve the poorest groups in their own development. The programme, whose funding is administered by the Central Bank, is geared to the poorest rural populations, and villagers establish their own list of priorities for which financing is being sought.

Lastly, there is the issue of whether it is in fact possible for rural-development projects conceived on an *ad hoc* basis to really achieve their stated objectives in the absence of a national strategy for rural development. Most governments do not possess such a strategy, though it is not uncommon to find that rural-development ministries or agencies exist in many countries. With most aid agencies, both multilateral and bilateral, showing increasing concern that their aid should be directed to the poorest populations, the need for such strategy becomes more urgent.

References

1 The more important agencies include the Ministries of Agriculture, Finance, Public Works, Water Resources, and Health and Assistance. Banks include the *Ejido* Bank, the Guarantee Loan Fund and the Agricultural Bank. Others are the Department of Agrarian Affairs and Colonisation and the National Corporation for Basic Marketing.

2 See Uma Lele, *The Design of Rural Development, Lessons from Africa*, a World Bank Research Publication, published for the World Bank by Johns Hopkins University Press, Baltimore, 1975.

PART II
Some principles of project formulation

5 Principles of project design and scale

We examined in Part I of the text the highly important role played by the project approach in the process of agricultural development in developing countries. Projects serve a variety of development objectives and the main project classes differ with respect to the demands they make on various factors of production, such as capital, land and labour, and also with regard to the distribution aspects of benefits generated.

With such a heterogeneous selection of projects and operating environments, the question arises of whether it is possible to distil the experience gained to date into a set of operational principles that would have practical value in their application, and thus improve the quality of projects being generated. The answer is clearly yes! Substantiating evidence derives, for example, from analyses of 'problem projects' of varying project classes in different countries. Failure can be traced to a surprisingly few common causes. For example, planners may have failed to evaluate properly the country's capacity for project implementation or the government may have failed to provide the matching local costs. Rate of acceptance by farmers was slow because of inadequate incentives. In this and the following chapter, we propose a far from exhaustive list of principles which mainly focus around the major observed defects in project preparation that affect subsequent performance.

Specificity of the objective

The project should have clearly specified objectives. Broad definitions – e.g. the objective is to raise agricultural production – are meaningless. A clear understanding of the objectives is important at four levels, namely, the country planning level; the project's management; the project's beneficiaries; and that of the agency/donor financing the project.

To illustrate this with a nutrition project, it is not enough to say that the project aims to improve the general level of nutrition in the country. The objective must be spelt out in terms of the target groups – for example, will the emphasis be on the undernourished in the urban areas and the migrants who are not fully absorbed into their new environment, or will the target group be the landless rural workers? This has implications for example, for the mode of production and channel of distribution of the food and also for pricing policy. The targets of change must be spelt out quantitatively, that is, within the identified vulnerable groups what percentage would be the target for the projects impact zone, and within what time-frame period – five years or ten years? Precisely what is to be changed in the nutritional pattern? This presumes that some consumption surveys have been carried out, and their daily total calorie intake of the target population averaged, say 1 500 calories per person. Nutritionists might have established that a desirable target for such a population, having regard to ethnic characteristics, climatic environment and normal workload requirements, should be, say 2 500 calories per person daily. The planning economist may then decide to scale down this target, to, say 2 000 calories on the social consideration that it is better to have 125 people fed slightly sub-optimally than 100 people fully fed.

Each decision taken sets the framework for a further derivative decision. Should the 500 calories gap per person be provided by increased production within the influence zone, or in another part of the country? What form would it take? The standard cereal staple, or should this be reinforced with pulses? This sets the problem in definite quantitative terms, for subsequent translation into an investment budget. However, in the process of projection formulation, it may have been discovered that the target groups suffered from severe intestinal worm infection. This, in turn, may have resulted from the absence of potable water supplies.

Thus, if the project's objective had been spelt out purely in terms of bringing the *per caput* calories of the group up to 2 000 calories and the project actions succeed in generating the targeted output, the project could be deemed successful, since it would have achieved its output levels. Yet, the real end purpose of the exercise – namely the physical welfare of the target group – has not been achieved. For the project to be a success, it must include project actions designed to remove the constraint to the effective utilisation of the additional calories arising from the project actions.

This is why it is important to spell out objectives. This, in itself, is a process of progressive refinement, but the importance of the clarification of objectives is that, it not only dictates levels and combinations of resources to be committed to the project, but also forces the project planner to think hard as to what he is really trying to achieve.

It is fundamental in setting objectives to recognise that the objective, once set, shapes the project actions. This is, however, much easier said than done because it is rare that a project has one single clear-cut objective.

The toughest decision arises where straight economic or production considerations must be balanced against other equally demanding considerations, such as income distribution and poverty redressal, or employment, or where slow-gestating projects must compete with resources for rapid gestation projects. In the design of a rice-production project, the project planners' ratings for different objectives, e.g. employment, income distribution and speed of development, will help to determine the choice of technology and the organisation of production. A weighting in favour of income distribution would normally imply a less concentrated project area, a larger number of beneficiaries and a lower level of capital per beneficiary, with attendant problems of project implementation and supervision.

A government, committed to the eradication of rural poverty, would have, at the initial project screening stage, deliberately to select those projects whose benefits would accrue to the poorer small farmers and the landless. This objective could well conflict, however, with that of achieving aggregate food-production increases in the short run.

Where the creation of new employment is an overriding objective, the planner would seek to maximise the use of labour per unit of capital spent. This has strong implications for the choice of technology and the role of mechanisation. For example, in a land-settlement project in the Terai of Nepal, which required the prior felling of marketable timber in areas selected for settlement, the government expressed a clear and firm intention that the operation should be as labour-intensive as possible. This in turn had to be balanced against the need to speed up the logging operation in the effort to relieve the pressure of the settlers for land. Thus in designing the project, power saws were proposed for the felling operation, but the construction of logging roads primarily would be done by hand.

Unfortunately planners are always faced with another dilemma. Since many international lending agencies finance primarily the foreign exchange content of the project, and maintain a discretionary position about local currency cost financing, the planner who wants to ensure a large foreign loan for the government, or the project, is, to say the least, tempted to inject mechanisation or equipment into the project costs at the expense of a greater use of local resources, mainly labour.

Finally, the clear specification of objectives is important for determining the content and depth of feasibility studies required during the preparation stages of the project. Where a large amount of costly and time-consuming engineering studies is entailed, it is important for both the lending institution and the government to ensure that their definitions of the project's objectives fully coincide at an early stage in the project-preparation process.

Flexibility in project concept and design

The project's targets and underlying strategy should not serve as straightjackets but should be subject to continuous review and modification as the project is implemented.

The planning process of governments in their national development has been undergoing considerable modification. In effect, what we observe increasingly today is the 'rolling plan' – that is, the five-year plan is treated as a perspective against the background of which, the annual plan is reviewed, revised and modified in a manner based upon successes, failures and setbacks in early years. This reflects the increasing realisation of uncertainties about planning projections five to six years ahead.

This development is relevant for project planning. Most international lending agencies are finding that there is a disturbing gap between commitments and disbursements, in their lending for agricultural projects. In other words, project implementation and execution are not proceeding at the rate envisaged at the project planning and appraisal stages. Whatever the multiplicity of reasons that may account for this situation, it seems reasonable to assume that a hard, probing look is required at one basic question – namely, has the project planning team paid adequate attention to the problems that would confront the project implementation team?

In practically all classes of projects we are faced with the question of uncertainty. Will the allocation made by the project planner for, say, a citrus component in an agricultural credit-lending programme be taken up by the citrus growers?

What are the possible precautionary and corrective measures that can be taken in light of marked uncertainty? Obviously the project planner can employ such tools as risk analysis, and sensitivity analysis to changes in demand, cost prices, project phasing, etc. However, he can do much more than that. He can build in flexibility to the project's design and concept, especially in those cases where the project is not subject to the constraints of technical and/or economic indivisibilities. Since most projects involve a mix of components, the analyst should identify which ones can be economically viable on their own account, if excessive delays arise during project implementation.

To illustrate this, in the Indonesian seeds project, the modular principle was employed. A module comprised a seed-processing plant with stated capacity, and the surrounding outgrowers provided the required numbers of hectares of rice-seed crops to service the plant. This permitted flexibility with respect to phasing of the project and of output levels.

Corrective measures are those that would be taken after the project becomes operational. One possibility is a review of the project at about, say, 18 to 24 months after it is operational, employing the data developed from the control system discussed in Chapter 6. The objectives of this review would be as follows:

1 To identify and assess the problems confronted during implementation by the project's management.
2 To review the basic assumptions of the project design and assess these against actual achievement.
3 To confirm or modify, as necessary, the project's objectives, targets, and basic strategy.
4 To assess whether a redefinition of the project is required. If such is the case, then it may be necessary to produce new cost estimates for the revised project, and also new financial and economic analyses.
5 To spell out the implications of these changes for the project's management and, of course, the government and the lending institution.

It is especially important that the project's management contribute to and participate fully in this review, which might also include the original project planners where possible. This would enable the project's implementation team to share in the planning process, in which they may never have been involved in the first place.

Obviously, these proposals have implications for the donor country and/or financing institution and the country because in effect we suggest that the project – scale, content and design – be not regarded as firm and definitive until the findings of the review mission are studied. The financing requirements of the project would thus be regarded as firm for the first two years, with the revised cost estimates from the project's review providing the basis for financing decisions affecting the remaining life of the project. Rural development projects with their varied components, might be the most appropriate with which to experiment before assessing the merits and pitfalls of this approach.

While this whole approach would make greater demands on staff time, the costs would be small in relation to the contribution it could make towards resolving the problems arising from painfully slow disbursements. However, this approach would also reduce the need for time-consuming preparation work.

Reconciling conflicts of interest

Where the project generates conflict of interest among components of the power structure, a firm commitment by government is indispensable for the project's success.

Every project has some implications for and impact on the existing power structure in the country, precisely because the project introduces a new flow of financial resources into the system, intended to change the production pattern or the ownership structure of the factors of production. The various power groups (e.g. *latifundists* versus *minifundists*), will attempt to capture as much of the flow of new resources as possible, or at least try to neutralise the flow of new resources so that they, at least, retain their privileged position.

The power groups, through their command of wealth and/or land, either hold the political reins directly or exert considerable political influence. Thus, it is naive for the planner to assume that the same type of economic growth is desired by all sections of the society. This is true for many Latin American countries where the average income of groups stratified by income varies widely, with the *per caput* income of the entrepreneurial group being 20 to 40 times larger than that of the vast majority of rural workers. The large land-owners would like to perpetuate the present situation here, where abundant supplies of cheap labour prevail.

Apart from the power situation at the national level, there are other types of sectional interests, both governmental and non-governmental. At the governmental level, Thailand provides a good example of the problems of project formulation. There are, for example, credit and marketing cooperatives, voluntary groups working with either the Agricultural Cooperative Bank or the Bank of Bangkok, the Farmers' Groups with the Rice Department, the Farmers' Clubs with the Agricultural Department, the Land Improvement Cooperatives and the Land Hire and Purchase Cooperatives of the Ministry of National Development, the Common Irrigation Organisation with specially recruited agronomists in the Royal Irrigation Department, and the People's Irrigation Associations in the Ministry of the Interior.

This plethora of government agencies and types of farmers' organisation poses an organisational nightmare for agricultural projects involving multidisciplinary approaches. Each agency jealously guards its authority and assesses the project's impact and benefits not on national considerations, but from a purely departmental standpoint. The planner thus has to be conscious of 'who's who' in the hierarchy and attempt to balance the diversity of interests. The creation of new autonomous agencies, that would provide greater latitude for project implementation, is thus strongly fought, as this inevitably implies some erosion of the authority sphere of the existing groups.

A further example of the problem of conflicting interests is the case in Indonesia where the launching of a seed project involving a new seed stream came in direct competition with the long-established village seed farms producing poor quality seed. This posed a threat to the financial interests of the involved groups. Furthermore, even within the Ministry of Agriculture itself the government-owned seed stations provided an important source of income to the extension-service staff who operated the stations.

Because of this it is especially important that the planner, whether local or foreign, try to approach project formulation with a balanced perspective. Quite frequently, experts associated with agencies – international and bilateral – try to impose their own viewpoints on a country. This could arise from their own one-sided evaluation of the problem or a desire to promote the viewpoint of their agency or the particular ministry to which they are attached. Sometimes their viewpoints arise from a superiority complex developed from their advisory role, where they feel that 'the locals do not know what is best for themselves'.

It is, therefore, important for project planners, especially those on a brief mission to a country, to interview the widest range of groups or interests possible with constant probing questions, in order to establish possible pitfalls or constraints to the project that may not be apparent, especially in the absence of a good knowledge of the country.[1]

From the foregoing, it is seen that it is difficult for the project planner to design projects with reasonable prospects of success. Hirschman[2] aptly points out the dilemma facing the project planner. In any situation the planner will be faced with negative attributes, e.g. lack of skills, fixed attitudes, etc. Thus the planner who operates on the basis that certain negative attributes of the *status quo* must be taken for granted as inevitable and unchangeable, could miss important opportunities for effecting positive changes in these attributes. If the planner takes risks and assumes that certain changes can be effected, and if these changes do not in fact occur and the project is in jeopardy then the planner would be accused of 'ignoring local circumstances, traditions and socio-political structure and of incorrigible naivety and lack of realism in general'. Hirschman thus proposes that the art of project design consists in escaping from this dilemma.

Whatever the situation, the project planner must recognise that government's commitment to the project is the major determinant of the prospects for success. Commitment implies two things – first, that the government fully accepts and understands the project's objectives, and second, that it is ready to allocate to the project the required resources, such as finance, personnel, and facilitating services. A lack of commitment may arise, for example where government's own priorities may not coincide with those of the donor country and/or international lending agency. For example, political considerations may weigh heavily in government's plans for a road network in contrast to the purely economic considerations of an external financing agency. The planner thus cannot assume that an identity of interests prevails and project design explicitly must take account of this. The country may also feel that the donor/lending institution is being too tough in the conditions for aid and may deliberately move slowly at the project's implementation stage.

A first requirement for ensuring the country's commitment is the active involvement of government in the selection and initial design of the project. This is possibly the only way to ensure that the government attaches high priority to the project. The term government must be interpreted in the broadest terms, to embrace planning ministries or units; the main implementing agencies, both at federal and state levels where these exist.

Adequacy of incentives

The project must ensure that the package of incentives it offers is adequate to generate the required responses on the part of the beneficiary.

The role of incentives in determining the response reactions of farmers to investment options is far from understood by most project planners. In project-preparation work, judgements relating to incentives tend to be rather mechanistic and lack reality because they are not based on a full understanding of the total environment in which the project will have to work.

Incentives may be defined, in this context, as a measure or group of measures, deriving from the project actions, deliberately designed to induce the project's target groups to assume the risks associated in bringing new and larger resource flows into production.

The key features of this definition are that with the project in operation, the farmer will, himself, be faced with new levels of investment, calling for a higher level of managerial skills on his part, e.g. proper use of irrigation water. This change from a traditional to a modernised agriculture introduced by the project confronts the farmer with a new risk element, for which he should somehow be compensated or protected.

The proper assessments of this risk, coupled with the level and nature of the 'incentive package', are important elements affecting project performance during the implementation stage. Typically in project work, the project's incentive contribution is taken to be the profitability of the farmer's investment. Farm budgets are derived and financial rates of return are worked out for project beneficiaries. A further important assumption in project formulation, required for assessing the project's economic benefits, is the rate at which new participants will be brought under the project's influence.

The indicator of net farmer income, however, does not always adequately measure the incentive effect of the project's investments on the beneficiary since the farmer takes other factors into account in his decision-making process. These considerations may be strictly of an economic nature but could also be social and attitudinal.

Factors of an economic nature include the following: the net benefits to the farmer estimated in the farm budgets may fail to reflect the loss of income to the farmer from work forgone in non-agricultural activities. Furthermore, the project's actions could constitute a threat to the historic economic relationships prevailing between the farmer and the intermediary, and his usual creditors. This is especially the case where traditional credit arrangements, including consumer credit, are linked to crop marketing. Thus, a credit project, limited to the provision of only production credit, might not constitute sufficient of an incentive to enable the farmer to undertake the risks entailed in the technological package being promoted by the project. A further major consideration is the absence of crop-protection insurance schemes in most countries. The farmer, therefore, may experience genuine fears that he cannot meet his credit obligations, now at considerably higher levels because of the project, in difficult crop years.

Also, of major economic significance is the question of who is the real beneficiary from the project's investments. In the absence of proper monitoring and control systems, very little is known about the distributional impact of such investments. However, prevailing tenancy systems are known to exert disincentive effects where tough land-rental conditions or share cropping prevail.

Social and attitudinal factors also enter the farmer's decision-making process. There may exist generally unfavourable attitudes towards agriculture. In the West Indies, for example, there is a massive exodus of young people away from agriculture, because agriculture is regarded as an inferior occupation, and the returns from farming are not such as to induce possible new entrants into agriculture to remain permanently on the land. In many Asian countries, projects involving double-cropping of land may pose conflicts where the farming community might normally use such time in the off-season for religious festivals, family visits and non-farm activities.

The role of incentives takes on particular importance in capital-intensive type projects, undertaken with public resources, but requiring complementary farm investment, if the production potential is to be realised. For large irrigation projects the whole question of response is critically important. One wonders for example, in a country like Thailand, where enormous investments have taken place in irrigation development with results far below those expected, whether the whole direction and scope of irrigation activities might not have been significantly affected if the question of farmer response to incentives had been more closely studied.

An earlier diagnosis of the problem could, for example, have permitted a better understanding of the factors that make farmers decide to undertake double-cropping, the rate of development of which is a major assumption in determining the project's economic rate of return. Furthermore, an understanding of the role of incentives would provide project planners with much clearer guidelines as to exactly how far the irrigation development should have been designed to bring water to the individual farm turn-outs. Thus at the project-planning stage, project components could well have included the provision of a tertiary network and even included land development. In short, the objectives of specific irrigation projects would have been more clearly spelled out, and less emphasis given to the more visible investments, such as dams and main canals.

Forms of incentives

An incentive package may take various shapes. First, on the cost side, purchased farm inputs may be subsidised; a common practice for fertilisers. Improved seed is another case. Agricultural implements and machinery are in many countries allowed in duty-free. Technical advice and other government support services invariably involve governments, absorbing at least a part of the costs. Farmers may also receive subsidised credit at rates considerably below the prevailing interest rates for other economic sectors in the country.

Another widely used approach on the cost side is for the government to pass on to the beneficiary only a part of the total investment costs of the project. This is common for capital-intensive type projects, especially irrigation, where water-use charges may reflect only the operation and maintenance costs of the system, with government or the irrigation authority absorbing all the capital construction costs. Reclamation and salinity control projects are further examples. In land-settlement projects, it is not uncommon for the government to bear the full costs of the physical and social infrastructure (roads, schools, etc.) with the settlers being entirely responsible for the on-farm development. Generally, lending agencies and donors tend to take a poor view of subsidies because they tend to distort proper allocation of resources.

The above cases represent clearcut, quantifiable subsidies, intended to exert an incentive effect on the project participants. A second class of cases provides the incentive of higher yields and increased net incomes through the adoption of a stated set of practices, including for example, the introduction of purchased inputs – seeds, fertilisers, pesticides, and mechanisation. This package could also include changes in the traditional farming system, e.g. introduction of double cropping, introducing a livestock component, introduction of a cash crop, e.g. cotton among subsistence farmers.

Finally, there is the incentive approach through pricing of the finished product.

Incentives and attitudes towards innovation

Myrdal[3] examined these problems in the context of the impact of the institutional structure on labour utilisation in south Asian agriculture. Because the bulk of south Asia's indigenous agriculturists have only peripheral contacts with the market economy, their primary economic objective remains that of meeting subsistence requirements. The share-cropping system serves as a powerful disincentive since the fact that rent varies, not with net return, but with gross output, means that the system has a strong built-in deterrent to intensified cultivation. The landlord himself has no incentive to invest, since he can get satisfactory returns without doing so.

Myrdal stresses the fact that technical experts fail to recognise the force of the reluctance of cultivators to change their work and their life, as technological change always implies. Promising technological innovations call for intensive programmes of training and education.

Mellor[4] points out that a traditional agriculture tends to be dominated by an attitude which emphasises survival and maintenance of position rather than improvement and advancement of production. This arises because of the high risk represented by innovation arising from the lack of systematic research or research on the overall system of farming and testing, and because the farmers' low income will not absorb a sharp drop before reaching minimal requirements for subsistence. He cites the following requisites for motivation to change: desire for increased material welfare; expectations that change will in fact increase wealth; and the expectation on the part of the innovator that he will himself participate in the net increase in wealth which accompanies successful innovation.

De Wilde[5] concludes from his analysis of the factors affecting receptivity and incentives to change in a rural society in Africa, that failures in agricultural development are largely traceable to inadequate knowledge or insufficient appreciation of the relevance of all the socio-economic factors that determine the proper approach to the farmer and that condition the farmer's response and receptivity to change.

He emphasises the decisive character of striking income increases if the innovation is to be accepted. Almost everywhere it was found that the prospect of modest increases in output had not stimulated much interest. The evidence indicates that the incentives required may vary considerably for differing types of projects. Thus, new irrigation or organised land-colonisation schemes may well have to offer significantly greater incentives to potential beneficiaries than those needed to induce acceptance of innovations on land where farmers are already established and are accustomed to live. Resettlement, for example, requires severing established social ties, adopting unfamiliar methods of production and accepting a degree of discipline under the supervision of a strange management.

Implications for project preparation

There is much scope for improving the quality of project preparation as it relates to the aspect of risks and incentives. Projects might be ranked, for instance, according to three levels or risks – high, medium, low. For example, high-risk projects might involve those with a new and untested technology, embracing for instance, new pasture species, new seed varieties, new breeds of livestock, new insect and disease contol techniques. Colonisation and transmigration projects, for instance, involve new environments, not only physical, in terms of soils and climate, but social, and possibly cultural and ethnic relationships also. The periodic incidence of natural disasters in the project area, e.g. droughts, floods and hurricanes, might render the project's investments subject to unduly high levels of risk. High risk might also arise from the nature of the demand for the product, especially perishables, such as fish, flowers and vegetables.

It is important, therefore, not only to assess the level of the risk, though this may not be exactly quantifiable, but also to identify the source of risk. Project design should then incorporate measures to reduce the impact of risk on the beneficiaries. To illustrate this, the use of collective security in a credit scheme reduces the risk element to individual farmers. The introduction of adaptive research in the project's actions would permit greater confidence in the technological package being promoted, especially if accompanied by demonstration blocks on farmers' fields. In fact, the project's size might be scaled down where it is clear that high risks exist.

The preparation of farmers in anticipation of assuming these higher-risk levels, should be a part of the project's design. This could be done by better informing farmers what the project is all about, and their expected role and contribution. This might be combined with socio-economic surveys designed to determine the possible response reactions of farmers to the project's proposals.

Having assessed the sources and intensity of the risk, it remains to assess the rewards the project offers in helping farmers to overcome these risks. Obviously, where the risks are high, the reward should be greater. Possible project tools to cope with this include interest rates to the beneficiary, grace periods and repayment terms and shares of total project costs borne by the beneficiary. The deliberate choice of beneficiary for the project's actions, such as small or landless farmers, is a powerful incentive that can be built into the project's design. Furthermore, the project planner might recognise that the removal of a disincentive can itself be a powerful incentive. A common example is the lack of foreign exchange availabilities for obtaining spare parts and supplies for agricultural equipment in many developing countries. Ensuring that the government will in fact make available the foreign exchange for, say, a pump irrigation scheme, constitutes in itself a strong incentive for farmers to enter such a project. Thus the incentive package needs not only comprise a financially attractive element, but should include all measures to reinforce the farmer's sense of security as a participant in the project.

Prompt identification and resolution of issues

At the very earliest stages of project identification, key issues should be identified and the mechanism set in motion to resolve issues.

A project always involves a variety of issues which should, in the main, be resolved before the target date for project implementation. Because the steps involved in bringing a project to a stage that can be called 'bankable' are sometimes tedious, time-consuming and costly in terms of finance and manpower, the implementation of later stages of project preparation should depend on an assessment of prospects for resolution of these issues. In fact, the ability to spell out these issues, confront them and thrash them out with the responsible authorities, lies at the very core of project preparation. In contrast, the investigation and data collection required in most projects may not generally involve any highly developed skills.

A clearcut identification of the issues is the first task facing the project analyst. An issue only merits attention if it materially affects the probability of success or failure in reaching the project's objectives. Viewed in that context, an issue can be seen as a 'bottleneck'.

It is helpful to know how the decision-making process, relevant to a particular issue, operates in the framework of the country. What are the standings of the respective involved parties in the power hierarchy and what types of decisions have to be referred to higher government levels? In other words, who is competent to speak for the government on a particular matter under adjudication? This is especially important where the issue involved crosses clear jurisdictional lines and contrasting viewpoints have to be accommodated.

In the latter case, which is very frequently encountered, the project analyst has to do a thorough 'staff job'. He must convince by the weight of his arguments but he must be prepared to make concessions employing the strategy of the least common denominator. At lower levels, departmental or even divisional considerations are apt to motivate the stand taken. Authority lines are jealously guarded, and it is important, in some societies, not to lose face or even to appear to lose face. These ministerial or departmental motives must not be lightly brushed aside. They may represent genuine concern on the part of the ministry about, say, the project's impact on their budgetary situation. At the higher government levels, especially the non-operational ones, like the national planning board or department, the concern is to ensure that the project proposal coincides with the country's national objectives. No doubt, sometimes, other considerations exist – the country may be anxious to secure a loan from an international financial agency so as to impress on its citizens its credit-worthiness abroad and its ability to obtain development loans. Meeting the spending targets of the national development plan may also be an important consideration.

What are these issues? They normally cover the entire range of a project's dimension, covered in later chapters. Briefly they may be organisational and managerial; technical or social; may relate to the project's size and general concept; or cover such matters as international bidding practices.

The importance of identifying these issues at an early stage and of resolving them cannot be stressed too much. Sometimes these issues are so difficult that they cannot be resolved through the whole stage of project preparation. Where the project is to be financed by an international lending agency, the issue may have to be resolved at the time of actual loan negotiations. Even then, occasionally, the issue may remain unresolved, and an already fully prepared and appraised project remains in suspense until a mutually satisfactory position is hammered out.

More extreme cases arise, however, when a government apparently accepts the lending institution's position on an issue at the stage of loan negotiations, signs the loan documents but attempts later to circumvent the requirements for compliance on an issue with which it was in basic disagreement.

We give below some illustrative samples, taken at random, of the various types of issues that arise during project preparation.

Organisation and management These are probably the most common issues and sometimes very difficult to resolve. Examples of the questions that arise are: should an existing national agricultural credit institution which has by accepted standards (e.g. loan disbursements, collections, etc.) proven itself incompetent, be used as the channel or implementing agency for a credit project, or should a new agency be created for the project? At the level of the project's management, the issue frequently arises as to whether an expatriate, selected for technical assistance to the project, can, in fact, have executive responsibility and exercise line authority over local staff. Many countries are increasingly opposed to such arrangements.

Economic issues An example would be the pricing policy being employed by the government for a particular product, important in the project's output mix. For example, the price of beef may be controlled to keep consumer prices low, but the levels may be such as to discourage beef farmers from expanding their production operations or otherwise modernising their operations, e.g. with improved pastures.

Financial issues Examples are many and varied and include to what extent farmers should be charged the capital construction costs of an irrigation system? Should a loan package to a farmer cover not only his purchased farm inputs but a share of his living expenses? Should the foreign exchange risk of the loan be absorbed by the government or the primary borrower or should it be passed on to the farmers? Where the working capital component of a project is large (e.g. an abattoir project involving regular purchases of cattle for slaughter), should this be included in the package for financing by an international agency?

Technical issues These refer largely to the extent and reliability of the existing stock of knowledge in the country and its adequacy as the basis for launching the project. Examples include: in an underground water irrigation project, is sufficient known about the characteristics of the aquifers to permit the large-scale sinking of tubewells? In a fisheries project, is there sufficient data on fishery resources to warrant an investment project involving purchase of more fishing boats, trawlers, etc., and installing on-shore refrigeration facilities?

Issues relating to scale of project (discussed below) are not uncommon. This is especially so where the middle-income developing countries are not interested in obtaining mini-loans from international financial agencies.

Project-scale determination

The project's scale is primarily a function of demand for the project's proposed output, of the country's overall resource situation and its absorptive and implementation capacity.

Scale itself, as a concept, has varying sides, but only has meaning when related to a particular time-frame in terms of workload, cost and output. The physical consideration of scale relates to geographic coverage of the project's main impact area; to the extent of physical project activities involved, e.g. the size of the irrigation structures and command area; or the amount of land to be clear-felled and plantations area to be established; and the number of farmers who will receive supporting services, and other such physically quantifiable features. The other important dimension measuring scale is the project's cost expressed as the level of financial and skilled human resources devoted to it.

Some of the more important determinants of project scale are discussed below.

The country's overall economic situation This is probably the most significant factor. Obviously what is large for one country may be small for another country, e.g. Somalia, or Chad, versus Brazil or Mexico. An analysis of the country's main economic variables, e.g. absolute level of GNP; the historic GNP growth rates, the balance of payments situation; the nature and structure of the country's outstanding debt; provides fairly reliable guidelines as to whether the project is likely to prove burdensome to the country or not.

Budgetary and financial allocations to the particular sub-sector These are important because they reflect not only the government's priorities in the particular field but provide a clue about the level of resources that the involved ministries and departments are typically accustomed to handle – though this may not imply that they cannot handle more resources efficiently. The scale of project proposals must, somehow, be examined not only against the background of the current public resource flows, but also must take account of private-sector flows where these are substantial.

The country's experience in a particular industry This is crucially important – obviously where a livestock industry is well established in a country, and its share of value of output is important in the country's gross agricultural product, then it should normally qualify for a larger project than one in which livestock development is a new industry, that has not reached sizeable proportions. The same applies to crop-production projects, where cultural practices are well known, and the constraints are lack of finance for further growth rather than a lack of technology. This is especially true, where there is an aggressive entrepreneurial group in the society, ready to take advantage of new market development.

Technical and economic indivisibilities These can dictate minimum project scale. On the technical side, an irrigation project may call for a certain minimum size of dam, and a minimum length of primary canal to serve the selected project area. Under such circumstances, there is a technical solution that is indivisible, and, therefore, sets the lower end of the project's scale. The Olmos project in Peru is an example, where the proposed tunnel through the Andes would involve a minimum length determined by nature. No benefits arise from a 99 per cent completed tunnel.

Economic indivisibilities stem from the basic economic nature of the operation which, in turn, sets a certain minimal physical base or output level to the project. For example, the use of a solvent-extraction process for groundnut oil manufacture can imply a higher minimum threshold of output than, say, for a plant employing mechanical presses. The situation is again demonstrated in forestry projects with an industrial component. To illustrate this, an investment in forest plantations totalling only, say, US$10 million might be quite adequate to support a chemical pulp mill costing, US$120 million. Chemical pulp mills with an annual pulp capacity of about 185 000 tonnes (air-dried) lie at the lower end of the scale of economic-sized mills currently produced by manufacturers.

Marketing and price prospects These are fundamental in arriving at project scale. The analysis should clearly differentiate between countries in a surplus or deficit situation with respect to a particular product. Where a large deficit exists in the domestic market for a particular commodity, and such demand can fairly accurately be measured, e.g. by import data (assuming that this data correctly reflects demand in the absence of controls, or strong import taxes) then the project analyst has a fairly reliable guide for setting the upper limits of his project. On the other hand, rice is not a homogeneous product and therefore total imports cannot be equated automatically with import-substitution opportunities. Where taste and quality factors come into play, these must be taken into account. The resistance of consumers in Indonesia to eating the rice from new high-yielding varieties of IR 8 series, in the initial years of introduction, is a case in point.

The same applies to industrial crops, especially cotton, where ecologically the country may only be capable of producing, say, short staple varieties, and there may be only limited possibility for substitution of the medium and longer staple varieties.

The other important element in import substitution is the question of price. The issue here is where a country is a high-cost producer, relative to the internationally traded prices of the particular commodity, is it justifiable to aim at self-sufficiency at all costs? There is no ready-made answer to the question. To some degree, it depends on the insurance premium that the country is ready to pay to guarantee its basic food supply. This usually is a political decision, but the project analyst must be ready to demonstrate the costs of this course of action to the country.

Where the project's output is predominantly export-oriented, the decisions are not so clearcut. Obviously, the country's competitive advantage in costs of production plays a significant role. Unless the country is ready to subsidise exports (e.g. through manipulating foreign exchange rate to benefit exporters), then the options are reasonably clear. Where a country's production of a particular commodity is only very marginal with respect to world trade in the commodity, and where no binding commodity agreement limits export quotas, then the country may well justify expanded production of the crop especially where the alternatives for agricultural diversification are limited.

Availability of managerial and technical skills These set the probable upper limits of the project. Even if the calibre of the top managerial ranks is adequate, the problem is invariably one of staffing at the middle and lower management, and foreman levels. The superstructure may exist but the infrastructure is missing.

Even where consultants have done an excellent job with, say, installing an irrigation network, it is the local extension staff – so frequently absent – who much teach the farmers proper water use and get them to utilise fully the installed facilities. The gap can only be filled by training programmes and ensuring satisfactory working conditions for the local staff. The situation calls for special attention in the poorest countries which lack both the capacity to prepare and implement projects.

Availability of labour This aspect deserves study not only for considerations relating to physical works to be constructed by the project, but is particularly relevant in assumptions about the availability of labour supply to cope with the seasonal labour requirements at the farm level. Particular attention must be paid to peak labour requirements at critical periods, e.g. harvesting, transplanting, etc. The farmer's non-farm cash income from other economic activities should be taken into account.

References

1 In one country, the writer asked a leading industrialist for his views on the problems of the banana industry. He replied as follows: 'Well, you know, other banana growing countries grow edible bananas. Here in this country we grow *political* bananas.' It was the most precise definition possible of the problems.
2 Albert O. Hirschman, *Development Projects Observed*, The Brookings Institute, Washington, D.C., 1967, pp. 130–32.
3 Gunnar Myrdal, *Asian Drama – An Enquiry into the Poverty of Nations*, a Twentieth Century Fund Study, Vol. III, pp. 1064–1069, Penguin Books, 1968.
4 John W. Mellor, *The Economics of Agricultural Development*, Cornell University Press, Ithaca, New York, 1966, pp. 244–65.
5 John C. de Wilde, *et al*., *Agricultural Development in Tropical Africa*, Vol.I, Johns Hopkins University Press, Baltimore, 1967, pp. 45–70.

6 Principles of control and evaluation

Built-in controls

The initial project design should include a built-in control system to monitor quickly the gap between established goals or objectives and actual performance so as to provide a guide for action by management.

Until very recently, the virtual disregard of the need to control or monitor projects constituted a severe defect in project preparation work. The very considerable effort that goes into planning the project – both conceptually and in manpower terms – is not at all matched by considerations about planning for a control system. In fact, one should think of 'planning' and 'control' as Siamese twins. If greater attention were paid to 'controls', it would force the project planner to provide greater clarity in the definition and planning of the objectives. Conversely, the more ill-defined the project's objectives are, the more difficult it is to design and operate a control system.

A control or monitoring system essentially involves a reporting system covering key variables that permits the project's management to identify quickly the areas where slippage is occurring, and how critical these slippages are to attainment of the project's objectives, so that corrective measures can be taken. Thus, the real purpose is to provide the project's management with the tools it requires to carry out its managerial function of control. The main characteristics of a good control system include the following.

1 The system should be simple to operate, and should be so designed as to meet the specific requirements of the project.
2 It should concentrate on key variables to be measured.
3 It should provide speedy answers, though approximate, to the critical determinants of project performance.
4 It should provide management with some leads as to what the corrective measures should be and how they should be introduced.
5 It should attempt to evaluate both elements which lend themselves to quantitative measurements and those that do not, such as attitudes of beneficiaries, which are important to the project's success.
6 It should highlight the precise source or sources of any problem, and assess this against the basic assumption employed at the stage of project planning.

The question may well be asked: Why should such stress be laid upon the need of control systems, when donor countries and multilateral financing agencies do in fact carry out supervision to varying degrees? Indeed, periodic field visits are undertaken and the government is alerted about the problems encountered in the project's implementation, with proposals for corrective measures required to get the project back on the planned course. These measures are far ranging and may cover such diverse features as personnel appointments, pricing policy, procurement of goods and services, and all the other possible issues that arise from a project.

However, an overriding consideration of the supervision process is to ensure that disbursement of approved loans takes place as scheduled, and is undertaken for those project actions for which loan funds were approved. The objectives of supervision, which can be compared with the external audit undertaken by the business firm, must be clearly differentiated from those of a control system which we earlier defined as a tool for management in its day-to-day operations of the project. The two can, in fact, be complementary, and the control system should provide data on the basis of which the supervision function could be better executed.

The need for control systems is much greater with the trend towards projects with greater social considerations, such as rural development projects, where increased agricultural production is only one of a project's different objectives. Increased rural incomes, and income distribution aspects come into greater focus, and it should be established whether the target groups are in fact receiving the expected benefits. The reduction of emigration to the cities also may be a consideration. The objective is to assess the project's performance as it evolves during implementation, so that management can influence the course of events rather than having a *post facto* analysis, which though interesting, may be too late to benefit the particular project.

A control system should serve two basic purposes for the management. The design of the system should differentiate between the data requirements relating to the execution of project works on the one hand, and the data requirements for assessing the impact of project actions on achievement of the stated objectives. All projects – especially those with a large infrastructure component – should have, at the design and planning stage, a simple critical path analysis that should alert the management about the major potential sources of possible problems. Problem projects, for example, show two main symptoms after the start of implementation, namely time overruns and cost overruns, with the former being much more serious than the latter. Apart from this, there is also the question of quality of the project works themselves.

Time overruns, for example, may result because the project's management is not fully aware of the government's procedure for, and the time required in, selection of contractors and consultants. Equipment may arrive at the project site, but the on-farm development work (e.g. land levelling) cannot proceed at the planned rate, because of inadequate ability of local trained staff to operate and service the equipment. In contrast, cost overruns may result from underestimates of the unit quantities involved (e.g. number of cubic metres to be excavated). In recent times, worldwide inflation has compounded the problem. Thus, the control system should guide management as to what are the key variables to be watched closely, and what options management has, to influence the rate of execution of project works during the construction period.

With regard to the achievement of stated objectives, we take as an illustration a project whose major thrust is the introduction of a new technology for raising production, e.g. improved seed, with a recommended package of cultural practices. The control system designed at the time of project preparation should be able to provide answers to such key questions as: What is the rate of acceptance of the new technology? To what degree is the recommended package adopted, e.g. fertiliser levels, seed rates, etc.? What are the major factors determining acceptance of the technology, in the project area or areas, e.g. farmer attitudes, input/output ratios in physical and value terms; effectiveness of demonstration and extension practices? What is the direct impact of the new technology with respect to increase in yield and production, gross and net farm incomes, labour utilisation, farm-gate prices? What are the indirect benefits or side effects of the project? For example, many farmers outside of the project area may procure seed from farmers within the project area but may not use anything approaching the recommended level of package practices. Above all, are the project's benefits in fact going to the projected target groups?

For each specific major item to be measured it is important to decide early, how this is to be measured. Thus, rate of acceptance can be measured both by total hectarage and number of farmers, over a stated time period, e.g. crop season, cropping year. Important questions also arise here, e.g. how long does the average farmer take to convert his whole farming area to the new technology?

How would the project's management employ such data in practice for the corrective measures proposed? We can take one such illustration. Let us suppose that the control system established that farmers are using on the average only 50 to 60 per cent of the recommended level of fertilisers. The next step would be to establish the reasons for this. We can envisage several possibilities. For example, the recommended package may have been developed by researchers at a central agricultural station, and its appropriateness for the soil types in the project area may not have been adequately tested.

Having identified the problem, management has to decide what action should be taken. For example, the project could arrange for demonstration and testing of various fertiliser levels. An intensified drainage programme may be indicated. The essential feature of all this is that the control system should reveal deficiencies and provide the basis for managerial action.

This whole aspect of designing control systems appropriate to a particular project in a particular country, is especially well suited to being undertaken by local research and development institutions as they possess intimate knowledge of the country. Lending institutions and aid-donors can do much to foster such an approach, which directly involves the country's own nationals in the development process.

Implementation of monitoring systems

The lending institutions have recently been paying much more attention to the monitoring aspects of the projects they finance. The International Fund for Agricultural Development has, for all the projects it has financed so far, introduced some covenants requiring the borrower to undertake some monitoring activities. This is especially important for IFAD because its lending policies require that the projects it finances should not have a negative impact with regard to income distribution, on the projects' beneficiaries. Simply expressed, this means that the rich farmers in the project area should not be made richer while the poorer farmers become relatively poorer.

The World Bank, however, is rapidly developing project-related experience in this field, and about 80 per cent of the agriculture and rural development projects it financed in fiscal 1977, had some component for monitoring and evaluation. In contrast, five years ago, while about a third of the projects it then financed had some reference to monitoring and evaluation, the financial provisions for this in projects was minimal. By 1977, the costs of monitoring and evaluation included in agriculture and rural development projects financed that year, totalled over US$20·8 million or roughly 1·7 per cent of the total base cost of projects with monitoring and evaluation components separately costed. The average cost per project in that year approximated US$1·0 million.

The World Bank conceives monitoring as the timely gathering of information on project inputs, outputs and complementary activities that are critical to the attainment of the objectives of the project. This should be a part of a broader *management* information system, including for example, general and cost-accounting system, internal auditing and managerial reporting. The measurable project objectives to be monitored include the following.

Project inputs These are the resource quantities necessary for a project such as capital, manpower and technology required to irrigate a certain number of hectares of land.

Project outputs These are the physical outcome of the project inputs such as the number of hectares actually irrigated; amount of credit provided, etc.

Project effects These are the outcome of the use made of the project outputs such as the crop yield per hectare of land irrigated.

Project impact These are the changes in living standards resulting from the project's effects, such as an increase in income and consumption, improved diets, increased local participation in planning.

Evaluation, in contrast, is the comparison of actual project effects and impacts against the established plans. *On-going evaluation* is the analysis that occurs during project implementation, while *Expost evaluation* is the analysis occurring after project completion and this is more relevant for policy making regarding future projects.

As an illustration, we discuss below some salient features of the monitoring arrangements for a rural-development project financed by the World Bank in Cameroon. The project's major components comprised institutional development, agricultural production, marketing and processing, social infrastructure and feeder roads. The project beneficiaries would involve about 16 500 farm families with a *per capita* income of US$85. The implementing agency is *Zones d'Actions Prioritaires Integrées* (ZAPI).

The types of information to be collected for monitoring would be as follows:

1 The major inputs such as financial, training, equipment, supplies and marketing that the project must deliver at the local level.
2 The activities and their duration as compared to the original plan.
3 The direct results or outputs of the various activities foreseen, such as number of kilometres of roads constructed, water supplies constructed, etc.
4 The key assumptions relating to project implementation which may require changes in the implementation strategy.

On-going evaluation would deal with the following types of information: income creation and distribution; satisfaction of basic human needs (health, food, housing); effectiveness of institution-building (cooperative structure); the degree of farmers' participation; and, behavioural and attitudinal changes. Thus evaluation would analyse the relevance of the implemented actions and their effectiveness in promoting rural development.

In designing the monitoring and evaluation system, the following potential constraints were recognised: the capacity of the project's management to absorb information; the human resource base to undertake the monitoring activity; the available financial resources since the borrower bears the cost of monitoring; the timeliness of data for management; and the availability of data-processing equipment.

The staffing requirements consisted of a chief evaluator supported by evaluators who would give the critical variables identified at the start of the project, data processors, and enumerators who would do data collection at the farmer level.

Technical assistance was proposed whereby consultants would assist in structuring the monitoring and evaluation framework, training of managers and evaluators; conducting a baseline survey; performing in-depth farm/village analyses; carrying out special topic surveys, and analysing the information collected.

Replicability

Where a first-phase project is to be followed by subsequent phases, the objectives and standards of the former must be replicable in the latter.

Many projects, at the stage of their conception, are intended as only the beginning of a larger development programme to be implemented over a much longer time-frame. Yet in the preparation of the project's objectives and design, quite often little or no attention is paid to the implications of this for the subsequent phases.

While not claiming that all types of projects are replicable, many types of agricultural projects do have that potential. A central research facility for a country covering the country's major crops is not replicable. Nor is an agricultural college that meets the country's training requirements for graduates in the visible future. However, the following types of projects are usually replicable: area development including rural-development projects; agricultural credit; land colonisation and settlement; crop establishment; adaptive research; input distribution or minimum package programmes; livestock and fisheries projects.

Thus the lending criteria of the main development institutions and aid-donors should more explicitly focus on the question of replicability of the projects they finance. The client countries can themselves, do much to intensify their efforts on the rural poverty front, by selecting and submitting for financing those projects which meet the tests of replicability. This is even more crucial in poor food-deficit countries, where poverty is a national objective.

There are a few very simple indices that planners could readily develop as guidelines in decisions as to whether projects meet the tests of reproducibility. The most important is the development cost per beneficiary. While estimates of the number of beneficiaries may be prone to error, especially in rural-development projects, the merit of this index is still beyond question. Other indices are development costs per unit of land, e.g. per hectare costs. Apart from financial costs, other indices, such as the use of scarce managerial and supervisory skills, are relevant.

Land-settlement projects in Nepal provide a good illustration of the types of problems to be considered in respect of reproducibility. The Nepalese Government began in the mid-1960s, with technical assistance from Israel, a programme for controlled land settlement in the Terai. Up until 1972, when the government sought the assistance of FAO and the World Bank, the programme had been extremely modest, with some 1 500 settlers having been assigned plots of land. The effort so far, pales into insignificance when viewed against the enormity of the task to be undertaken. It is estimated, for example, that the Terai could accommodate some 240 000 settlers – and settlement of the Terai is the obvious and most logical outlet to relieve the enormous land and population pressure in the hill areas of Nepal. Request for World Bank assistance in settlement of the Terai was intended to accelerate the process. The World Bank project, as prepared by FAO, was intended as a first phase of a long-term programme of controlled settlement of the Terai.

We discuss below some of the more significant points which should be considered in a first-phase project within the perspective of the total development of the Terai. These points would generally apply to most settlement projects.

Development costs per settler This consideration is of overriding importance. Where the problem is large, and the country resources are limited, there is no justification for a 'show-case' approach in the first phase, which the country cannot afford to replicate in subsequent phases. In fact, the decision arrived at here, about the amount of capital to be invested per settler, should set the whole tone of the development in ensuing phases.

Farm sizes Decision-making with regard to farm size cannot be related merely to the proposed development in the first phase. The appropriate decision must be viewed within the context of the total problem – namely, what is the extent of the land resources capable of supporting a sustained agriculture and what is the rough order of magnitude of the settlement problem facing the country? This must be related therefore to man/land ratios in the main agricultural areas of the country.

Other considerations come into play. The objective of settling as many farmers as possible in the long-run must be balanced against the objective of ensuring that excessive fragmentation does not result in the creation of non-viable farming units, incapable of supporting the farm family of average size. This, in turn, raises the question of farm income. Other points to consider in determining farm size are the possibilities for irrigated agriculture, the soil fertility, crop rotations proposed, family labour availability and mechanisation (including draft animals).

Farm income In every land-settlement project, the question arises of what target should be set for the net farm income of settlers. This is not a simple question and is usually not left alone to the economists as this has strong political implications. The more general considerations that come into play are the country's *per caput* income level, disparities between urban groups and the rural groups, and the prevailing income levels in the area or region where the project is being introduced, and in the area of origin of the settlers.

Crop rotations The cropping rotation proposed for the first-phase area should broadly be applicable to the areas in subsequent project phases. For example, if a maize–pulse (gram) rotation is proposed, these crops should be suitable for the subsequent phases.

Extent of social and support services The social services provided by the project such as schools, medical facilities, drinking-water supplies, etc. should be such, that they can be reproduced in follow-up phases. Other support services, such as agricultural extension should be provided at a density that makes sense within the long-term perspective of the overall national development.

The price of land The price which settlers in a first phase project would have to pay for their land, should be arrived at by much broader considerations. The price set should more closely reflect the total demand and supply assuming that free market forces were at work. The land price should take the farmers' repayment capacity into account. It should certainly not be set at levels that are conducive to speculation.

The settler's contribution The guidelines and practices adopted for the settler's contribution in the first phase should be equally applicable to subsequent phases. If, for example, the settlement's road infrastructure costs are fully borne by the project, without any attempts to recover this from the settlers, then this policy would have to be followed in subsequent phases.

The foregoing is intended to show that the level of resources entering the first-phase development must be such as, not only to ensure the project's success, but also not to jeopardise the country's ability to replicate similar development for the benefit of other beneficiaries.

While attention has been focused here on land-settlement projects, the issues raised are especially pertinent for rural-development projects. Can the poorer countries, where the need for such projects is most urgent, afford politically and economically to take on expensive first-phase projects which do not meet the tests of reproducibility? The project analyst should, therefore, be fully aware of the dimensions of the problem.

Adequacy of operating capital

The operating capital and foreign exchange requirements must be guaranteed so as to ensure that the physical facilities created by the fixed investment can be utilised to their maximum capacity and maintained over time.

The foregoing seems so obvious as to hardly warrant the status of a principle. Examples of failure arising from this are, however, encountered so frequently that it is clear that at the project planning stage, inadequate attention was paid to this aspect, or it was presumed that somehow the working capital requirements would be forthcoming. UNDP/FAO projects in a host of countries suffer from this problem. Ministries of Agriculture in several countries cannot properly execute their tasks because of this.

In fact, the project planner can get a realistic feel of the country's problems and attitudes by observing how well the country's stock of capital is maintained. Do projects function well on the whole? Are roads well maintained or allowed to degenerate? Are there frequent breakdowns of public services, e.g. transport, water, electricity, etc. (excluding the human factor of strikes)? Fixed assets just cannot produce without the lubricant of operating capital.

Although this principle applies to all classes of projects, it is more pertinent for those projects with a trading or a processing component. Raw material resources must be purchased largely on a cash basis; they must be processed, and the inventory of finished goods financed. It is not uncommon to find that the full magnitude of funds for inventory financing has been underestimated. In fact, even in small processing units such as groundnut processing, copra processing, etc., the working capital requirements may be as much as four to six times as large as the fixed investment costs.

The importance of this problem looms larger when viewed in the context of local cost versus foreign exchange costs. Within the working capital requirements of, say, a rice-processing plant, such items as fuel and spares, might account largely for the foreign exchange component. However, the largest cost components would be raw material (paddy) purchases and the wage bill. Since international financial institutions have generally been guided by the foreign exchange component of the project, in arriving at the extent of the loan, the country is generally faced with meeting the bulk of the local cost requirements of the project, and inadequate budgetary provisions may be made for working capital.

With rampant worldwide inflation pushing up interest rates, the problem of working capital, especially inventory financing, requires even greater attention on the part of project planners. The question of liquidity takes on greater significance. In livestock enterprises, for example, especially beef-cattle fattening operations, the working capital requirements are tremendous, and their availability must be guaranteed.

The project planner, when in doubt, can justify the case for a smaller project, to ensure that project's operating capital requirements are satisfied. Whatever measures are taken, the best guarantee, however, is to confirm that the government is in fact committed to the project. The existence of a loan document does not suffice.

Evaluation of project's induced costs

The decision to undertake a project should be based on an evaluation of all identifiable project costs, and not only direct costs.

It is a valid criticism of most project planners to say that they tend to accentuate the positive, the assets side of the project's balance sheet tending to be more heavily weighted than the liabilities side. This obviously contributes to a more favourable economic rate of return for the project, and thereby enhances the project's chances for approval. Certainly many instances exist where a project would not have been undertaken if the principle of 'full costing' had been employed. At least the project would have been modified somehow in recognition of these costs.

The costs referred to here include such elements as the impact on employment, the income losses of affected groups, the production losses arising from the drainage of limited human resources to meet the project needs; and induced foreign exchange leakages. To some measure, the problem of cost measurement arises because there is a built-in tendency to limit the cost consideration to what is happening within the project area but not outside the area. Yet, it is precisely outside of the project area that the adverse impact of the project's costs are most likely to be felt. For example, finance may be concentrated in one area to the exclusion of others.

The most frequently encountered cases refer to managerial and technical skills. The donor and/or multilateral financing agency may insist on the very best calibre of managers and technicians, which are almost invariably in short supply. The problem is considerably aggravated where the project represents a completely new departure from normal operations and a new entity must be staffed with personnel from existing institutions.

In justifying the benefits from a rice-production project, yields in the project area might be claimed to rise from, say, 2·0 to 3·0 tonnes paddy because of the new inputs package levels and because of the intensive technical guidance that farmers would receive from the intensification of agricultural extension services. This argument concedes, therefore, that extension is a powerful element in the raising of productivity and production. The corollary, therefore, would be that the draining away of staff from the agricultural extension services should result in production decreases *outside* of the project area. This, we maintain, is a cost to the project, particularly where the project does not stand the cost of replacement of the seconded staff. In any event, there would be a time lag involved while the replacements were trained to the level of their predecessors.

The more acute problem arises at the top management levels of the project. It can be argued logically that top management in developing countries has an extremely high opportunity cost because of the scarcity factor, and this is probably considerably higher than the other factors of production such as capital and labour. Yet, whereas the economic analysis is based on assumptions about the opportunity cost of capital in the country, and also considers other factors, e.g. scarce foreign exchange, generally, the analysis is quite moot about the scarcest factor, management. Present analytical techniques for projects do not tackle this problem.

Apart from managerial and technical skills, other scarce elements come into play. When seed and/or planting materials are in very short supply, or breeding stock for a livestock programme are strictly limited, then the priorities for the project's needs require that non-project beneficiaries must forgo or defer their own development programmes until additional supplies are forthcoming. The same applies to other scarce resources, e.g. purchased farm supplies, pesticides, spare parts for which only a very limited amount of foreign exchange may be available.

Projects with a large mechanisation component can also be problematical. Where the effects of such mechanisation are output-increasing, e.g. when heavy mechanical ploughs provide adequate tillage in heavy-textured soils, or facilitate timeliness of operations, then positive benefits can derive from mechanisation. Where, however, mechanisation is introduced in the project for cost-reduction purposes, then there may be costs to the country of absorbing any displaced labour, or alternatively of depriving the under-employed labour pool of finding an outlet for their labour within the project.

Finally, various projects have varying impacts on the foreign exchange leakages they induce. This is especially relevant in the Latin American context where the larger farmers go in for conspicuous consumption and their profits end up in vacations abroad, imports of luxury goods, and cash balances abroad. This is more important where the country's fiscal tools are not geared to tapping the surplus generated by the project for reinvestment in the country. One further consideration in projects geared towards medium and larger farmers, e.g. agricultural credit projects, is whether the project's loan funds result in an additional flow of resources for on-farm investment or whether such funds merely substitute the farmer's own resources, which are then channelled into higher-yielding investment, e.g. apartment houses for rent.

If such costs could be identified and quantified, they might reduce the attractiveness of financing this class of farmers on the basis of production considerations alone. In short, we make a plea for project planners to ask more searching questions about the true costs of the project, even though they may not lend themselves to quantification with sophisticated measuring tools.

PART III

Project design and analysis – main considerations

7 The technical basis

The experience gained by agencies involved in financing agricultural projects suggests that the major dimensions of the project are: technical, institutional, organisational and managerial, financial and commercial, economic and social. Although convenient to compartmentalise them for analytical and discussion purposes, project formulation requires, in practice, an appreciation of their interacting nature. For example, economic viability is largely a function of technical viability. The project stimulates a highly dynamic situation which is continuously evolving over its lifetime, in contrast to the pre-project situation of stagnation or minimal change.

Definition

All projects have or should have a technical basis which should be clearly defined. The technical assumptions shape important derivative assumptions such as degree of farmer-participation, production and farm incomes – in fact all the final coordinates that make or break the project.

An all-embracing definition of the technical basis is not easy, because of the vastly differing nature of agricultural projects. However, the following definition is proposed:

> A project's technical basis comprises that store of applicable knowledge about the physical and biological inter-relationships of the system, that permits prediction of responses, within acceptable margins of error, to changes in any important physical parameters, to the point where the outcome of an investment decision can be quantified.

In the majority of agricultural projects, we would be concerned with a physical system, comprising soils; crops and cropping systems; livestock, water systems, farm inputs (including labour and machinery); ecology, including climate. Except for weather (and even this is marginally susceptible to influence, e.g. cloud-seeding techniques) the other physical components, can be controlled in varying degrees. For example: the physical characteristics of heavy clay soils can be changed by heavy doses of organic manures. The crop-production potential of acid soils can be altered by appropriate lime applications. Naturally such changes have economic limits to their widespread application.

The aim of the technical analysis is to obtain sufficient knowledge of how the whole physical system would operate with those changes that would be introduced by the project, so that responses can be determined – responses being measured as crop-yields per unit area, beef and/or milk; timber increment/density, etc. In short, we are trying to establish production functions.

The data base is derived from various sources, including: an examination of existing experiences in and around the project area; the appraisal of work on research stations; the undertaking of feasibility and pre-investment studies; the judgement of experts in the field, and experience in other countries with roughly similar conditions.

Constraints to technological change

Some writers argue that technological reform can only be successfully introduced where required institutional changes have been introduced. Thus Myrdal states:

> The major theme of our analysis will be that the subject of technological reforms, designed primarily to increase the cultivated acreage and raise agricultural yields through variations in techniques and the input of capital, hinges largely on the extent of prior or least simultaneous institutional changes. In fact, the two types of agricultural reforms – the technological and the institutional – have tended to be compartmentalised in public discussion and policy formation, both during the colonial era and since independence.[1]

Myrdal's broad generalisation is, however, open to question. Institutional changes are slow and expensive to implement especially where these involve changing the structure of ownership of factors of production, e.g. land tenure. It may be more appropriate to say that projects can considerably increase aggregate production, but would have only limited or even negative income-distribution aspects within an existing deficient institutional framework. Thus, in the Indian sub-continent, the technological breakthrough, spearheaded by the high-yielding varieties, resulted in considerable increases in grain production without any marked structural changes. In fact, one is left to wonder whether drastic institutional changes should be introduced in the absence of other growth stimuli such as a strong technological breakthrough.

The absence or weakness of the research base is a serious constraint to raising production in many developing countries. De Wilde[2] points out that in the African context, several problems which limit the development of agricultural production have not yet been solved by agricultural research. Among the more important instances cited are: the general problem of conserving soil fertility where long-term bush fallow is no longer feasible, since no really satisfactory and definite substitute has yet been developed for bush fallow. There is still no answer to the unexplained failure of legumes to produce nodules and fix nitrogen under the conditions prevailing over most of tropical Africa. Furthermore in livestock production, there are no answers to the central problem of overgrazing on much of the rangeland when population increases.

Allan[3] in supporting this view, points out that no one can say with any certainty what these carrying capacities are, for there is little factual evidence and there is need for a great deal of experimental work on a wide variety of pasture types.

Yet it seems to be in the area of food-crop production that research has been most deficient as has been emphasised by Lele[4] in the African context. It is a fair generalisation to say that the industrial export crops such as oil palms, rubber, cacao, sugar cane, coffee and cotton, have generally received a far larger share of the research dollar than food crops. This is certainly true for the West Indies, where the highly organised commercial sugar industry does research on sugar cane. Cacao and bananas have received considerable attention, whereas it is only recently that the Faculty of Agriculture of the University of the West Indies has begun paying attention to the region's basic foodstuffs, e.g. yams and other root crops. Apart from rice and corn, the lack of research knowledge on the starchy foods, and such cereals as sorghum and millets, is appalling. Research required covers not only the production or yield aspects, but also nutritional aspects and simple processing methods.

It is frequently encountered, however, that research at the national-level institutions has not been adapted to the needs of the country's major ecological zones and is a constraint in project formulation. Projects with adaptive research components are increasingly important because of this gap.

Purposes of the technical analysis

The technical analysis comprises one key element in the overall feasibility studies required before the government or donor or aid-agency can decide on financing the project. It serves the following more important purposes.

To establish the extent of and capacity of a physical resource for development The exploitation of surface and underground water resources, of a river basin, fishery resources, standing timber, marshy or swampy land, national grazing stands, virgin land for agricultural development, drought prone areas, requires a prior technical evaluation of the capacity of the resource for sustained development.

To generate proposals for possible development alternatives and to justify the selected alternative There is invariably more than one alternative to exploiting a resource, and options must be identified, weighed, rejected or selected. The choice of a dam site and decisions relating to the height of the dam may have to take into account not only cost elements, but the sometimes competing uses of stored water for irrigation development, power generation, tourism development and security from potentially damaging floods. The alternative selected should also justify, for example, the method of construction, as this could well have a major impact on the employment and income-generation effects of the project during the construction period. Is the time gained by using expatriate contractors with masses of heavy equipment, worth the forgone opportunity of using a labour-intensive approach, that reduces unemployment levels? Is the resource being developed with a view to accommodating as many beneficiaries as possible? Thus, if the least cost alternative in monetary terms is selected, the costs to the country of other development priorities, e.g. employment and poverty redressal, should be known.

To establish the major physical components of the project The technical basis establishes the project's major components. In an irrigation project, a dam of a predetermined size would be built, and irrigation structures commanding a defined geographic area would be built. In a seeds project, processing plants would be established to a predetermined capacity, a stated hectarage of farm land would be developed for seed growing; seed testing laboratories would be established. Thus, the technical analysis establishes the hardware requirements of the project.

To establish the proposed development plan at the farm level On the basis of agronomic data, farm-income targets and market considerations, the cropping pattern for the project area would be developed, with *pro forma* development plans for representative types of farms. The package of agricultural practices is proposed, based on assumptions about technical input/output relationships, that is, simple production functions are established. The application of one kilogram of urea should result in a stated increase in production of paddy under the expected conditions of water management and control. All these farm-level assumptions are fed in to developing the farm budget, discussed later in Chapter 10.

To provide the basis for deriving the project's cost estimates In projects involving considerable civil engineering, work quantities have to be estimated, and unit costs determined. In land-settlement projects, costs of land clearing must be determined, whether this is to be mechanical or done by hand. At this stage also, the foreign exchange component of project costs must be established.

To establish the project's period and phasing The physical phasing of project works is important for several reasons. There may be interest costs to be borne by the borrower during the construction period, and this may have to be included sometimes in the loan for the project. With inflation, unrealistic proposed construction periods may pose serious financial threats to the project's completion at the scale envisaged. The phasing of project works also establishes the sequence and levels of flows of the project's benefits, such as the number of farmers brought in annually to an irrigation or a land-settlement project. These assumptions should not be mechanical, but should reflect the realities of the country's situation, especially the capacity of the project's management and the speed with which contractors and consultants can be brought into the stream of project activities.

To provide the basis for detailed engineering and design during the construction phase The preliminary designs, prepared at the feasibility stage to establish technical viability and cost estimates within acceptable margins of error, form the basis for detailed engineering and design required for the construction phase. They provide the basis on which the government, with agreement of the lending agency, defines the terms of reference for consultants and contractors, and on which tenders are issued and evaluated.

To supply the major inputs required for the economic analysis Economic data such as project costs, broken down by major works or activities to be undertaken, are provided for analysis. Investment costs are separated from operating and maintenance costs. Additionally, the physical output stream in units of production is estimated, so that project benefits can be quantified in value terms.
To provide a basis for formulating the project's organisation and management Ensuring that the major assets, created by the project do, in fact, generate expected production increases is the task of the project's management. Identifying and logically grouping the project's activities that derive from the technical proposals, are a first step in evolving the organisation and management structures. The allocation of project costs by major activity gives an indication of the priority that each grouping of activities should have in the mangement's structure and attention.

Burma: irrigation development

The proposed project located in central Burma, would involve three major components, first, the construction of a storage dam, located upstream from an existing diversion weir, which would provide water primarily for irrigation and limited electricity generation; second, irrigation development, which would comprise the rehabilitation of two existing irrigation systems in areas A and B, totalling 45 325 ha and the construction of new irrigation systems for two areas not presently served by irrigation and totalling 20 230 ha; third, agricultural development of the above, including the proposed cropping patterns, the support for seed production and agricultural research activities, and improved mechanisation services.

The following discussion is limited to the technical assumptions around which the proposed agricultural development of the areas was based, with specific reference to the cropping patterns. The description of the existing agricultural situation briefly covers soils, land utilisation, cropping patterns, crop yields and production, and existing research in the project area. This provides the basis on which the project proposal was drawn up.

Soils and land

Soils in the project area had been studied by the Land-Use Bureau and a soil survey team of the Irrigation Department. Soil-survey data covered the physical and chemical properties of the soils, including infiltration rates, and the groundwater table and salinity conditions. Soils were classified and mapped according to the Burmese classification, which was based on Russian soil-classification concepts. A brief description of soils in the project follows.

1 Meadow soils (gleysoils) – about 80 per cent of the rehabilitation area and 25 per cent of the new areas; these are brownish grey mottled clay soils which characterise the low-lying areas with high groundwater tables.

2 Dark compact savannah soils (vertisols) characterised by heavy texture with deep solum. Their heavy texture and compact nature precludes dry-season cultivation without irrigation.
3 Red-brown savannah soils (ferric luvisols) which are mostly well-drained sandy soils, with low moisture-retaining capacity.
4 Alluvial soils (fluvisols) which are characterised by well-drained coarse-textured recent alluvial deposits, with good internal permeability but low moisture-holding capacity.
5 Cinnamon soils (chronic luvisols) which are well-drained sandy loams that have developed on elevated alluvial plains. They are moderately well drained with a deep humus horizon.

Land classification – the suitability of land for crop production – was in accordance with Burmese classification standards which are mainly based on topography, soil texture and drainage conditions. Three main categories were identified as follows:

1 Paddy lands – these include the meadow soils mainly in the existing irrigation tracts, which have for a long time been cultivated to paddy. Because of the high groundwater table and imperfect natural drainage, they would require a surface drainage system for production of non-paddy crops in the dry season.
2 Upland and paddy lands intermixed – the soils of this land group include dark compact soils and meady cinnamon soils. These lands support non-paddy crops and some paddy and cotton rotation under irrigated conditions.
3 Uplands – these are broken down into *ya* lands which are the upland soils on highland, and *kaing* which are the alluvial upland soils along the rivers, and include the cinnamon soils, red-brown savannah soils and alluvial soils. Their coarse texture and high permeability and slightly undulating topography make them suitable for non-paddy crops.

Existing agricultural production and practices

The location of the project area in the dry monsoon zone of Burma, with an annual rainfall of 813 mm spread over seven months, is the main determinant of the existing land use and cropping patterns. Rainfall patterns are not evenly spread through the season and show strong yearly fluctuations. The predominant crops are paddy and cotton – the latter representing the results of a government-sponsored cotton programme in the area.

Table 7.1 shows the gross and net sown areas, the cropping intensity, yield and production, without the project and with the project at full development for the two existing irrigated areas A and B, and the non-irrigated areas C and D. In the existing irrigated areas, the cropping intensity, ranging from 130 to 140 per cent is determined by the availability of irrigation water from the existing irrigation systems and the special government cotton programme. Figure 7.1 shows the cropping calendar. Paddy is the predominant crop in the monsoon season, followed by cotton when irrigation water is available during the time of cotton planting in February/March. Double cropping of paddy – about 5 per cent in the irrigated areas – is minimal because available water is

TABLE 7.1 Burma: Irrigation project – Crop area and production

Crop	Net area (ha)	Cropping intensity (per cent)		Gross sown area ('000 ha)		Yield (kg/ha)		Production ('000 tonnes)		Incremental production ('000 tonnes)
		O[1]	W[2]	O	W	O	W	O	W	
Rehabilitation area A										
Local paddy		60	27	21·4	9·5	1 700	2 130	35·4	15·8	−19·6
HYV paddy		15	70	5·3	24·8	2 170	3 480	11·4	84·8	+73·4
Cotton (medium staple)		30	30	10·7	10·7	505	1 010	5·3	10·6	+ 5·3
Groundnuts (in shell)		1	4	0·4	1·5	561	1 010	0·2	1·5	+ 1·3
Sesame		17	8	6·0	2·9	123	252	0·7	0·9	+ 0·2
Pulses		5	1	1·8	0·4	336	650	0·6	0·3	− 0·3
Chillies		6	10	2·1	3·6	548	774	1·2	2·7	+ 1·5
Onion		6	10	2·1	3·6	4 000	5 051	8·5	17·7	+ 9·2
Total	35 600	140	160	49·8	57·0					
Rehabilitation area B										
Local paddy		49	25	4·8	2·5	1 700	2 130	7·9	5·1	− 2·8
HYV paddy		10	83	1·0	8·5	2 170	3 480	2·1	27·5	+25·4
Cotton (medium staple)		3	10	0·3	1·0	505	1 010	0·1	1·0	+ 0·9
Groundnuts		2	6	0·2	0·6	561	1 010	0·1	0·6	+ 0·5
Sesame		30	14	2·9	1·3	123	252	0·3	0·4	+ 0·1
Pulses		20	2	1·9	0·2	336	650	0·6	0·1	− 0·5
Chillies		8	10	0·8	1·0	548	774	0·4	0·7	+ 0·3
Onion		8	10	0·8	1·0	4 000	5 051	3·0	4·8	+ 1·8
Total	9 700	130	160	12·6	15·6					

New area C										
Rainfed paddy		5	–	0·4	–	1 100	–	0·4	–	– 0·4
Local paddy		–	20	–	1·7	–	2 130	–	3·5	+ 3·5
HYV paddy		–	57	–	4·8	–	3 480	–	16·5	+16·5
Cotton (local)		5	–	0·4	–	214	–	0·1	–	– 0·1
Cotton (medium staple)		–	28	–	2·4	–	1 010	–	2·4	+ 2·4
Groundnuts		2	14	0·2	1·2	314	1 010	0·1	1·2	+ 1·1
Sesame		20	18	1·7	1·6	62	252	0·1	0·5	+ 0·4
Pulses		33[4]	3	2·8	0·3	336	650	0·9	0·2	– 0·7
Chillies		8	10	0·7	0·8	225	774	0·2	0·5	+ 0·3
Onion		2	10	0·2	0·8	4 000	5 051	0·6	4·2	+ 3·6
Total	8 500	75	160	6·4	13·6					
New area D										
Rainfed paddy		5	–	0·6	–	1 100	–	0·5	–	– 0·6
Local paddy		–	17	–	2·0	–	2 130	–	4·2	+ 4·2
HYV paddy		–	37	–	4·4	–	3 480	–	14·9	+14·9
Cotton (local)		5	–	0·6	–	214	–	0·1	–	– 0·1
Cotton (medium staple)		–	47	–	5·5	–	1 010	–	5·4	+ 5·4
Groundnuts		3	17	0·4	2·0	314	1 010	0·2	2·0	+ 1·8
Sesame		18	18	2·1	2·1	62	252	0·1	0·6	+ 0·5
Pulses		27[4]	4	3·2	0·5	336	650	1·0	0·3	– 0·7
Chillies		1	10	0·1	1·2	225	774	0·1	0·9	+ 0·8
Onion		1	10	0·1	1·2	4 000	5 051	0·5	5·8	+ 5·3
Total	11 740	60	160	7·1	18·9					

[1] O = without project. Estimate based on crop data for A, B, C and D areas, provided by Land Records Offices
[2] W = with project overall intensity based on three cropping patterns
[3] Late monsoon season
[4] Areas under pulses include wheat, maize, sorghum, etc.

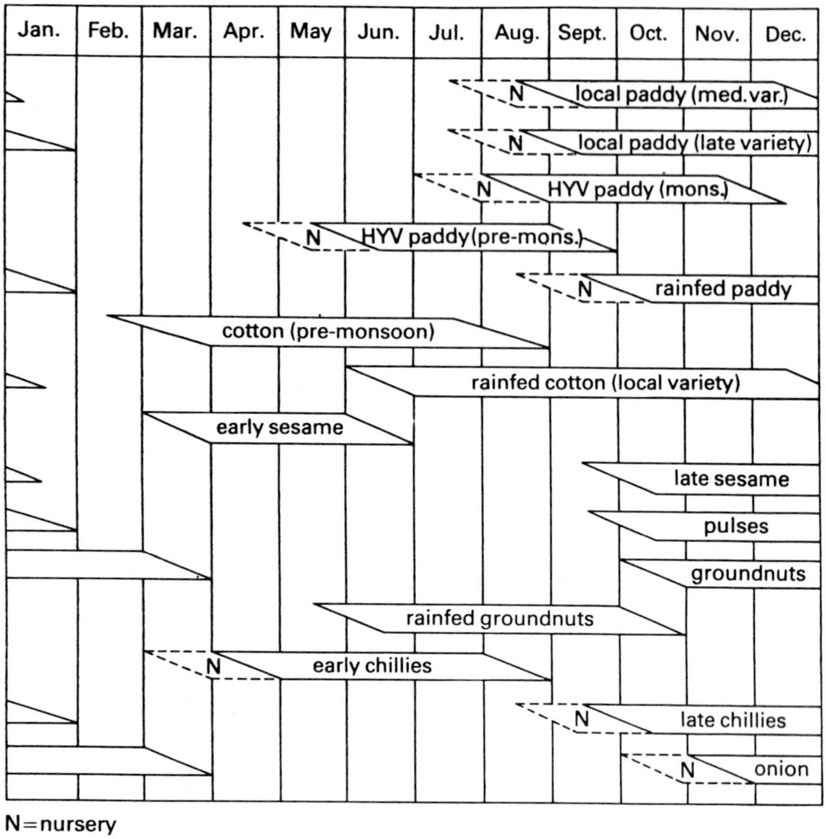

Figure 7.1
Burma: Irrigation project – Present cropping calendar.

allocated for cotton growing. The monsoon paddy is sometimes preceded by early sesame, chillies and vegetable crops. Onions planted in the late monsoon season (October) are sometimes followed by cotton. Chillies and onions are intensely cultivated under irrigated conditions in limited areas.

In the non-irrigation areas, C and D, where crop production is dictated by variable rainfall, the cropping intensity ranges from 60 to 75 per cent. Paddy, though a preferred crop, is limited to low-lying areas close to the river. Main crops are oil seeds – sesame and groundnuts and pulses. Some short-staple cotton is grown and other crops, in limited quantities, include chillies, onions, maize, wheat and sorghum. Almost a half of the total cultivated area is left in fallow. Some dairying is practised.

Paddy cultivation involves mainly the use of indica type local varieties, of which the medium ones have a growing period of 150–170 days, while the late varietal groups have a growing period of 170–200 days. High-yielding varieties (HYVs) are grown in about 20 per cent of the total paddy area. Yields are low – averaging 1 700 kg/ha for the local varieties, and 2 170 kg/ha for the HYVs (Table 7.1). Low yields arise because of the present inadequate on-farm water delivery and low levels of inputs, especially fertiliser, and the poor quality of HYV seeds. Yields on a research station in the project area exceeded 3 920 kg/ha with HYVs. Fertiliser use on paddy is limited to the HYVs and is well below recommended levels.

Cotton under irrigated conditions is grown on about 30 per cent of the areas. The varieties are of medium-staple American upland cotton. Planting occurs in February/March on land mechanically prepared by tractors from the Agricultural Mechanisation Department. The inadequacy of these tractor services results in delays in planting time and these spread out to mid-April, which implies harvesting under wet conditions. Poor seed quality requires seed rates of up to 40 kg/ha, and plant population stands are low. Irrigation water is applied two or three times, and spraying is done by hand at the beginning of the monsoon.

The low yields, averaging 500 kg/ha, derive from:

1 generally poor drainage in the area;
2 poor quality of seeds;
3 low plant population;
4 inadequate pest control;
5 inadequate irrigation water during the vegetable growth stage and excessive rain in the later growth stages;
6 inadequate selection of suitable areas for cotton;
7 lack of interest from farmers and incentives for cotton growing.

The proposed development

This took account of the fact that the project would introduce certain actions designed to remove the constraints to increased agricultural production in the project area. Briefly, these constraints are: inadequate water supply from the existing irrigation system, the lack of surface drainage and inadequate on-farm distribution. Additionally, the arrangements and levels of input requirements for the more important crops were deficient. This applied mainly to seed, fertilisers and pesticides. In the case of cotton, inadequate mechanisation services for land preparation prevented the timely planting of the cotton crop. In the irrigated areas, possiblities existed only for one monsoon crop.

The proposed cropping patterns were worked out for the three major land groups identified in the project area, and reflected the experience gained from the work of agricultural research stations in the project area. Soil survey findings indicated that soils in the existing irrigated areas were suitable for paddy cultivation, and for non-paddy or upland crops provided good drainage was possible. The project provided for improved drainage. In the new area to be brought under irrigation, soil surveys had established that about 60 per cent of the soils are suitable both for paddy and non-paddy crops, with the remaining 40 per cent being more suitable for non-paddy crops under irrigated conditions.

Thus, in formulating a new cropping pattern, an attempt was made to establish a balance between the physical potential for development, represented by the area's agro-climatic conditions and the project's activities in irrigation and drainage and the existing cropping patterns. In

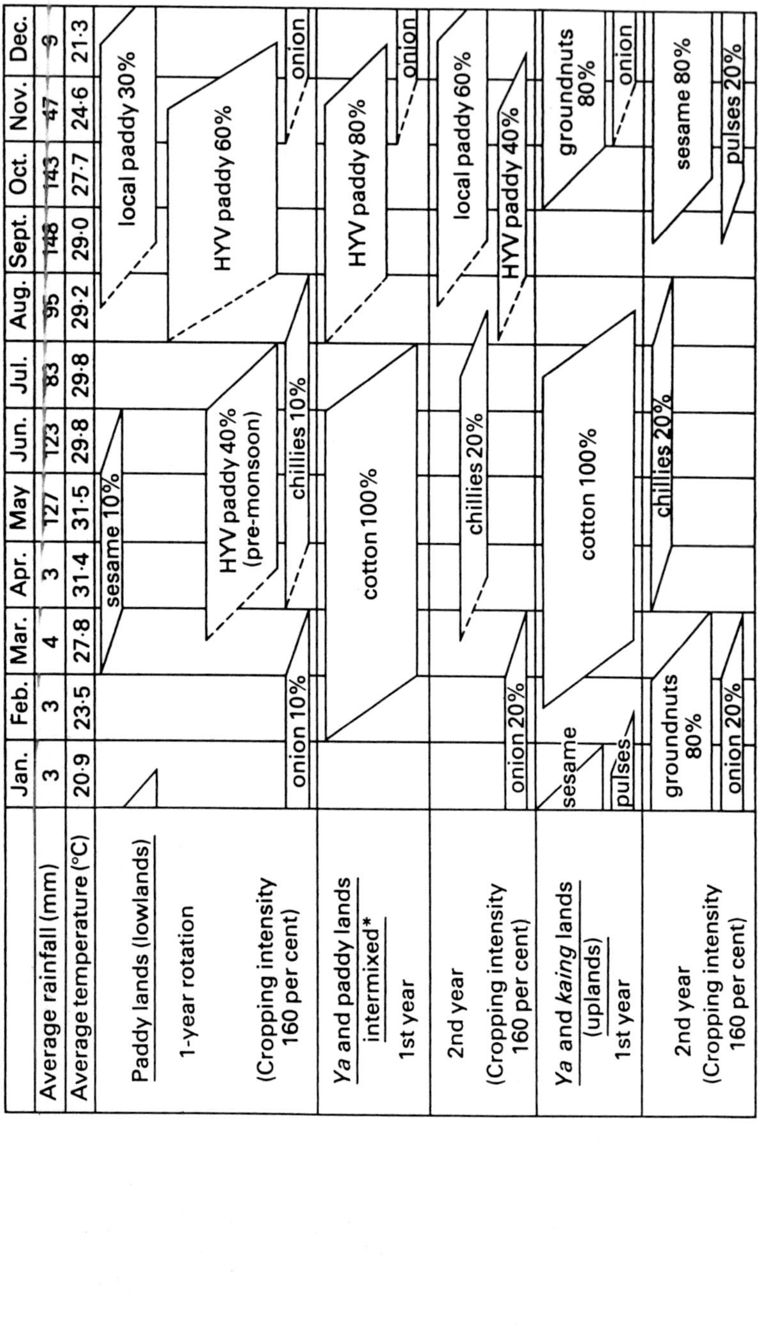

* Paddy lands in area A where the groundwater table is below 0·9 m, would follow this rotation in order to maintain the present cotton hectarage in area A in the same proportion in the future.

Figure 7.2 Burma: Irrigation project – Proposed crop-rotation pattern.

addition the pattern had to take into account the government's expressed wishes to secure reliable and increased cotton supplies to supply the cotton processing facilities already established at great cost, close to the project area and operating below the rated capacity. The farmers expressed wishes, based on interviews, were also taken into account. To illustrate this, it would have been desirable to change the planting time of cotton from the present February/March season to the late monsoon one of September/October, so that the cotton harvesting is undertaken under drier conditions. The reaction of farmers was negative because of their traditional paddy-orientation, and their reluctance to devote their paddy fields to cotton growing during the main paddy season. Futhermore, paddy cultivation was voluntary, while cotton cultivation was mandatory.

The proposed cropping pattern is shown graphically in Figure 7.2 and Table 7.1 which shows the projected situation as W – that is the 'with project' situation.

For the paddy lands, the cropping pattern would be basically a one-year rotation of the pre-monsoon paddy (40 per cent). According to farmers this would be the maximum manageable proportion for an average holding of a 2·8-ha paddy farm, taking account of labour requirements at the harvesting time, without introducing mechanical threshing and drying facilities. Sowing would be advanced to permit harvesting in July when the average total rainfall is 75–100 mm. The pre-monsoon paddy would be followed by monsoon paddy (90 per cent) of which HYVs would make up to 60 per cent, and local varieties 30 per cent. This provision for local varieties takes account of consumer taste preferences for farmers' home consumption. Other crops would include chillies, onions and sesame – 10 per cent each (Figure 7.2). Under this proposed rotation, the cropping intensity would rise from the present range of 130–140 per cent upwards to 160 per cent.

For the *ya* and paddy lands, a two-year rotation is proposed for pre-monsoon cotton and monsoon paddy. Thus the planting season for cotton remains unchanged (Figures 7.1 and 7.2). Pre-monsoon paddy is not proposed for this crop rotation, in order to avoid growing paddy on the next plot to cotton. Chillies and onions would also be grown on small areas. The overall cropping intensity would be 160 per cent of which 50 per cent would be cotton, 60 per cent for HYV paddy, 30 per cent for local paddy and 10 per cent each for chillies and cotton.

For the *ya* and *kaing* lands, the prosed two-year rotation would comprise: cotton, groundnuts, sesame, chillies, onions and pulses (Figure 7.2). Because of the permeability of these soils, no paddy is proposed for this land class. Groundnuts, a high-value crop, would be grown right after the pre-monsoon cotton in the first year, then followed by onions and chillies, sesame and pulses – the traditional crops of the *ya* and *kaing* lands. Overall cropping intensity would be 160 per cent of which 50 per cent for cotton, 40 per cent for groundnuts, 40 per cent for sesame and 10 per cent each for chillies, onions and pulses.

As far as the formulation of the proposed cropping patterns is concerned, the major changes introduced by the project are detailed below.

1 For the existing irrigated areas, modest increases in the cropping

intensity from the existing range of 130–140 per cent to 160 per cent. While progressive individual farmers in the project area considerably exceeded this target figure, the project's assumption of an average of 160 per cent seemed more realistic. The thrust of these changes revolved around paddy production, with minimal changes proposed for cotton. The increased water availability, stemming from the project's actions, would facilitate a considerable intensification of paddy production. This provided the opportunity for the widespread introduction of high-yielding varieties, which require much larger water supplies under controlled conditions, to achieve their potential, than is the case with the low-yielding varieties. Thus, in A, one of the existing irrigated areas, it is noted (Table 7.1) that the cropping intensity for HYV paddy would rise from an existing 15 per cent to 70 per cent, while the cropping intensity of the local paddy varieties would decrease from 60 to 27 per cent. This has strong implications for the yield and production estimates projected in Table 7.1 for all four sub-project areas. In fact, one can conclude that the introduction of high-yielding varieties, including adequate quantities of quality seed, comprises a major element of the project's technical basis at the farm-development level.

2 For the new areas to be brought under irrigation, the proposed cropping intensity would rise substantially from its present range of 60–75 per cent to 160 per cent. Main changes would be the introduction of paddy – especially HYVs – and substantial amounts of cotton production, reflecting both the soil capabilities and government policy. There would be a notable reduction of maize, wheat and sorghum.

The proposed intensification of cropping and the larger prominence of high-yielding paddy varieties and cotton , naturally call for higher levels of purchased farm inputs, if projected yields, stemming from the project's actions are to be realised. The requirements of urea, tri-superphosphate and potassium chloride were derived for the project area by using unit recommended figures on a crop basis for each of the sub-project areas (Table 7.2).

Research findings in the project area, and recommendations of the concerned government agency and prevailing practices in other countries with a similar agro-environmental situation, formed the basis for developing these proposals. Because of the need for cotton spraying to control insect pests – an important element in cotton yields in the area – a similar calculation was prepared to cover the pesticide requirements.

Two other proposed project actions were significant. First, because of the shortage of quality seed for the two main crops, paddy and cotton, an existing government farm would be fully strengthened with buildings, equipment and staff to provide the project's seed requirements. Arrangements for research and demonstration would be improved. Second, existing tractor services for field preparation for cotton would be considerably strengthened, and proper equipment for cotton spraying would be provided.

Based on the foregoing project actions, yield and production increases were derived and are shown in Table 7.1. Thus, the local paddy yield would show modest increases, rising from their present levels of

TABLE 7.2 **Burma: Irrigation project – Estimated fertiliser requirements**

Item	HYV paddy	Local paddy	Cotton	Groundnuts	Sesame	Pulses	Chillies	Onion	Total (tonnes)
Requirements (kg/ha)									
urea	80	67	78	34	23	23	101	101	
tri-superphosphate	67	34	39	45	34	34	101	101	
potassium chloride*	34	–	–	–	–	–	–	–	
Total requirements (tonnes)									
Area A									
gross sown area (ha)	(24 800)	(9 550)	(10 700)	(1 500)	(2 900)	(4 050)	(3 560)	(3 560)	
urea	4 380	630	825	50	65	10	350	350	6 660
tri-superphosphate	1 640	375	410	65	95	15	350	350	3 240
potassium chloride	820	–	–	–	–	–	–	–	820
Area B									
gross sown area	(8 050)	(2 460)	(1 010)	(570)	(1 330)	(160)	(970)	(970)	
urea	1 420	160	80	20	30	5	95	95	1 905
tri-superphosphate	530	80	40	25	45	5	95	95	915
potassium chloride	275	–	–	–	–	–	–	–	275
Area C									
gross sown area	(4 800)	(1 660)	(2 390)	(1 170)	(1 580)	(280)	(850)	(850)	
urea	850	110	180	40	35	5	85	85	1 390
tri-superphosphate	320	55	90	50	50	10	85	85	745
potassium chloride	160	–	–	–	–	–	–	–	160
Area D									
gross sown area	(4 370)	(2 020)	(5 460)	(2 020)	(2 070)	(490)	(1 170)	(1 170)	
urea	770	130	420	70	45	10	115	115	1 675
tri-superphosphate	290	65	210	90	70	15	115	115	970
potassium chloride	155	–	–	–	–	–	–	–	155

* Muriate of potash

1 700 kg/ha to 2 130 kg/ha. This reflects their general low-yield potential. The high-yielding varieties are projected to rise from present levels of 2 170 kg/ha to 3 470 kg/ha. This sixty per cent increase for the HYVs would arise from the special project actions to ensure quality seed in amounts required for the project area, much greater water availability and improved water distribution and control at the farm level. Additionally, the improved drainage facilities provided by the project would considerably enhance the performance of the HYVs.

In the case of cotton, a doubling of yields is projected from the present low levels of 500 kg/ha, seed cotton, to 1 010 kg/ha (Table 7.1). This projection reflects the expected performance of the overall average cotton producer rather than what could be expected of the best farmers. In fact, it may be argued that the projected yield is low, having regard to farmers' performance in similar parts of the world, such as Pakistan. However, the proposed technical package for cotton would include improved seed, adequate water requirements, controlled planting time facilitated by improved mechanisation services, a larger plant population per hectare, improved pest-control practices, supportive research and technical expertise in cotton cultivation. With improved drainage, the interaction of all these proposed measures should permit a doubling of yields from such low levels.

It should be pointed out at this stage that the bundle of agricultural practices proposed in a project of this type, with heavy civil works component, can be critical to realising the economic benefits of the investment. The agricultural practices are also a key determinant of the financial rewards that participating project farmers can be expected to receive.

Mali – integrated rural development project

This example demonstrates the basic technical assumptions employed in the design of a rural development in Mali. The project was designed to accelerate the production of groundnuts and food-crops.

In order to reverse the country's declining groundnut production, the government, with French bilateral assistance, introduced *Operation Arachide* (OA) in 1967. The earliest efforts towards the operation were supported by a 25 per cent increase in produce prices and by 1969, OA initiated measures to improve farming practices and promote use of modern inputs. OA further promoted a functional literacy programme among farmers.

By 1972, OA's performance was already impressive. By that year's growing season, 52 000 OA farmers growing 99 000 ha of groundnuts produced 76 000 tonnes of groundnuts or about one-half of the country's production. In that season, half of the OA farmers sowed at the prescribed time, 30 per cent at the recommended density, 36 per cent applied fertilisers and 32 per cent used seed dressings on groundnuts and 20 per cent used seed dressings on millet and sorghum. After 5 years of OA operation, improved seed distribution increased 12 fold. Yield increased from 500 kg/ha in 1968 to 814 kg/ha in 1971, the area under OA supervision doubled, and national groundnut production increased from 81 000 tonnes in 1967 to 153 000 tonnes in 1972.

The project, now discussed, represented an intensification and extension of OA activities to new areas, and external financing was obtained to this end. Because the scope of activities was considerably broadened to include food-crop production the project was renamed *Operation Arachide et Cultures Vivrières* (OACV), while management responsibility remained with the previous OA Board of Directors and operational managers.

The objectives of the new project were as follows:

1 To intensify OACV activities in areas currently served by the programme.
2 To cover, not only groundnuts, but all major farming activities mainly the production of basic cereals and livestock.
3 To expand OACV services into new areas.
4 To supply farm inputs and equipment, on cash or credit, to participating farmers.

Supporting activities provided by the project included a programme for improvement of rural tracks; a functional literacy programme; an agricultural research programme involving agricultural diversification studies; improved groundnut varieties and soil surveys; improved medical and veterinary services; and a project-evaluation unit.

Target hectarages were established annually over the life of the project. Participating farmers would move through stages of improved practices on these areas. The farmers rate of absorption of these practices would depend mainly on their existing cultivation practices, receptiveness to advice, and credit-worthiness, as this would determine their use of improved inputs. All participants would grow millet and/or sorghum in rotation with groundnuts, and would receive technical advice on how to increase yields. The project benefited from the fact that OA had introduced a range of recommended practices over the past five years for groundnuts, including early and dense sowing, timely weeding, fungicidal seed dressing, fertiliser and the use of improved seeds.

The overall cropping pattern proposed under the project aimed at the following:

1 Self-sufficiency in food crops for the farm family.
2 Eliminating labour bottlenecks at times of land preparation, weeding and harvesting.
3 Permitting cereals to follow groundnuts in the rotation so that the cereals could benefit from the residual effect of fertiliser applied to the groundnuts.

A typical rotation for a farm using fertiliser would be as follows: year 1, groundnuts; year 2, millet/sorghum; year 3, groundnuts; year 4, millet/sorghum; and years 5 and 6, bare fallow.

Components of proposed technical package

Groundnuts The technical package recommended was expected to increase the basic yield by a factor of 2·4. The main components were:

Early sowing This ensures that the complete growing cycle falls within the rainy season, the optimum sowing dates being from the end of May to mid-June. Correct sowing time increased the basic yield of 500 kg/ha by a minimum of 20 per cent or 100 kg/ha. Every delay of one day after recommended planting time reduces potential yield by one per cent.

Spacing The traditional wide spacing with a plant population of 40 000 plants per ha (seed-rate of about 40 kg/ha), contributed to low yields. The optimum for the heavily branched types is a population of 90 000 to 111 000 plants per ha (seed-rate of about 150 kg/ha). Dense sowing improved the basic yield by a minimum of 40 per cent or 200 kg/ha.

Improved varieties Two varieties suited for differing ecological zones had been developed. The project's target was that 70 per cent of farmers in the project area would use seed of improved varieties in contrast to 49 per cent in the pre-project period. The use of improved seed would increase the basic yield by a minimum of 20 per cent or 100 kg/ha.

Seed dressing This protects the sown seed from attack by various pests, thus preventing a second sowing or having to fill gaps. Thus seed dressing with thioral at a rate of 100 g/kg of undecorticated seed nuts was recommended. The use of fungicide would increase the basic yield by a minimum of 10 per cent or 50 kg/ha.

Mineral fertiliser The recommended fertiliser – single superphosphate (21 per cent P_2O_5) applied at a rate of 65 kg/ha – was expected to increase basic yield by a minimum of 50 per cent or 250 kg/ha. Research on application for various soils and crop rotations would be carried out under the project's research component.

Cereals The main sources of yield increases would derive from the residual fertiliser effects of the rotation in which millet and sorghum follow groundnuts and seed dressing. The use of deep ploughing with ox-drawn implements, has shown evidence of beneficial effects in Senegal and in the OA areas. Table 7·3 shows the impact of the various practices on yield.

In order to work out the increases in production induced by the project, and required later for the purposes of economic analysis, it is necessary to establish first the rate of adoption of these techniques. Table 7.4 shows the projections of diffusion of modern techniques for

TABLE 7.3 Mali: Integrated rural development – Impact of recommended practices on yield

	Groundnuts		Millet/sorghum	
	per cent	kg/ha	per cent	kg/ha
Basic yield	100	500	100	600
Early sowing	20	100		
Correct density	40	200		
Use of fungicides	10	50	15	90
Improved seeds	20	100		
Fertiliser application and residual effect	50	250	25	150
Deep ploughing	–	–	30	175
Total	240	1 200	170	1 015

TABLE 7.4 Mali: Integrated rural development project – Area under groundnuts: diffusion of modern techniques in project area

	Pre-project		PY1		PY2*		PY3		PY4		PY5		PY6		PY7 onwards	
	ha	per cent	ha	per cent	ha	per cent	ha	per cent	ha	per cent	ha	per cent	ha	per cent	ha	per cent
Total area cropped	99 600	100	104 600	100	138 100	100	156 800	100	172 400	100	181 600	100	186 950	100	191 050	100
Timely sowing	49 900	50	57 500	55	62 100	45	78 400	50	94 800	55	108 900	60	112 100	60	114 600	60
Correct density	30 100	30	36 600	35	41 400	30	54 800	35	77 500	45	90 800	50	93 400	50	95 500	50
Using fungicides	32 350	32	43 900	42	55 200	40	70 500	45	86 200	50	99 800	55	102 800	55	105 000	55
Using improved seeds	49 000	49	60 600	58	75 900	55	94 000	60	120 600	70	127 100	70	130 800	70	133 700	70
Applying fertilisers	36 150	36	49 100	47	55 200	40	79 500	45	94 800	55	99 800	55	102 800	55	105 000	55
Ox-powered cultivation	2 400	2	4 350	4	7 450	5	13 200	8	20 950	12	30 400	17	35 000	19	46 300	24

* Decrease in percentages due to the addition of 26 000 ha of the San Segou sectors where no improved techniques are used.

TABLE 7.5 Mali: Integrated rural development project – Estimated production and yield of groundnuts in project area

	Production without improvement		Incremental output due to modern techniques										Total production and average yield under project	
			Timely sowing		Correct density		Use of fungicides		Use of improved seeds		Application of fertilisers			
Years	Total area cropped (ha)	Production[1] (tonnes)	Area (ha)	Production[2] (tonnes)	Area (ha)	Production[3] (tonnes)	Area (ha)	Production[4] (tonnes)	Area (ha)	Production[5] (tonnes)	Area (ha)	Production[6] (tonnes)	Total production (tonnes)	Average yield (kg/ha)
Pre-project	99 600	49 800	49 900	4 990	30 100	6 020	32 350	1 620	49 000	4 900	36 150	9 040	76 370	767
PY1	104 600	52 300	57 500	5 750	36 600	7 320	43 900	2 195	60 600	6 060	49 100	12 275	85 900	821
PY2	138 100	69 050	62 100	6 210	41 400	8 280	55 200	2 760	75 900	7 590	55 200	13 800	107 690	780
PY3	156 800	78 400	78 400	7 840	54 800	10 960	70 500	3 525	94 000	9 400	70 500	17 625	127 750	815
PY4	172 400	86 200	94 800	9 480	77 500	15 500	86 200	4 310	120 600	12 060	94 800	23 700	151 250	877
PY5	181 600	90 800	108 900	10 890	90 800	18 160	99 800	4 990	127 100	12 710	99 800	24 950	162 500	895
PY6	186 950	93 475	112 100	11 210	93 400	18 680	102 800	5 140	130 800	13 080	102 800	25 700	167 285	895
PY7	191 050	95 525	114 600	11 460	95 500	19 100	105 000	5 250	133 700	13 370	105 000	26 250	170 955	895

Basic yield (undecorticated): [1] 500 kg/ha.
Additional yields (undecorticated): [2] 100 kg/ha; [3] 200 kg/ha; [4] 50 kg/ha; [5] 100 kg/ha; [6] 250 kg/ha.

TABLE 7.6 Mali: Integrated rural development project – Incremental groundnut production induced directly by the project

	Pre-project	PY1	PY2	PY3	PY4	PY5	PY6	PY7 onwards
Area (ha)								
without-project[1]	99 600	103 700	135 200[2]	148 700[2]	156 400	159 000	159 000	159 000
with-project	–	104 600	138 100	156 800	172 400	181 600	186 950	191 050
Increment	–	900	2 900	8 100	16 000	22 600	27 950	32 050
Average yield (kg/ha undecorticated, rounded)								
without-project	767	756	697[3]	697[3]	670	667	667	667
with-project	–	821	780	815	877	895	895	895
Increment	–	65	83	118	207	228	228	228
Production (rounded)								
without-project	76 370	78 430	94 170	100 920	104 770	106 070	106 070	106 070
with-project	–	85 900	107 690	127 750	151 250	162 500	167 285	170 955
Increment	–	7 470	13 520	26 830	46 480	56 430	61 215	64 885

[1] Without-project situation assumes that some additional farmers will grow groundnuts under improved methods and some existing farmers now using improved methods would further increase their hectarage and productivity.
[2] Sharp increase in PY2 and PY3 reflects inclusion of Segou Region in project area in PY2 (26 000 ha) and Kenieba district in PY3 (6 650 ha).
[3] Average yield declines due to inclusion of Segou Region and Kenieba district with their low traditional and non-improved yields.

groundnuts into the project area. Two factors are at work, namely, increases in area, and the proportion of the cultivated area benefiting from each of the recommended practices. For example, while the cropped area would almost double, the percentage of the area benefiting from correct density would increase from 30 per cent in the pre-project period to 50 per cent by the fifth year of the project's life, shown in Table 7.4 as PY5.

The next step is to estimate the production and yield of groundnuts in the project area. Table 7.5 shows the production without improvement, the incremental output deriving from the recommended practices, and the total production and average yield under the project. The coefficients from Table 7.4 have been used to derive Table 7.5.

It will be noted that the initial average yield of 767 kg/ha applies to OA areas already benefiting from improved practices.

Finally, it is necessary to isolate the effects of the project, which must be assessed later against its costs. To do this, it is necessary to estimate what would have happened in the absence of the project, that is the without-project situation, to establish the incremental benefits arising from the project.

Table 7.6 shows that the incremental benefits of the project's production would arise from two sources. It is noted that the incremental production directly attributable to this project would, on an annual basis, be 64·885 tonnes (by the seventh year of the project) over the production in this pre-project period. This would arise from the project's increasing the cultivated area by 32 050 ha with an incremental yield of 228 kg/ha from the fifth year onwards.

References

1 Gunnar Myrdal, *Asian Drama – An Inquiry into the Poverty of Nations*, a Twentieth Century Fund Study, Vol. III, p.9, Penguin Books, 1968.

2 John C. de Wilde, *et al.*, *Agricultural Development in Tropical Africa*, Vol. II, The Case Studies, Johns Hopkins University Press, Baltimore, 1967.

3 W. Allan, *The African Husbandman*, Oliver and Boyd, Edinburgh, 1965, p.292.

4 Uma Lele, *The Design of Rural Development, Lessons from Africa*, a World Bank Research Publication, Johns Hopkins University Press, Baltimore and London, 1975, p. 181.

8 Institutional features

The institutional background of a project is a product of the country's historical evolution and provides the framework within which the project will have to operate. It would be unrealistic to suppose that an unlimited sweep of changes could be introduced by the project; the planners must tailor the project to the existing background.

A main task during project preparation is, therefore, to identify those components of the institutional framework that have a bearing on the project. In most countries, however, the planner is likely to encounter a large number of institutions involved in agriculture and this intensifies problems of coordination in projects, whose activities cut across various lines of authority. Among the more important possible groupings or elements determining this background are: government institutions; quasi-public bodies, corporations and marketing boards; land-tenure systems; banking and credit institutions; farmer organisations and cooperatives; water and water rights and covering legislation; and religious customs and practices and social mores as they determine village authority.

We briefly discuss and give illustrative examples of how the more important ones are liable to impinge on a project. Although the institutional background has the greatest implications for the organisation and management proposed for a project, it also affects project formulation in various ways (e.g. through inducing modifications in the project concept and strategy).

The way in which the institutional background influences distribution of the gains from the project's investments is probably its most important feature. The project has to operate in the framework of the country's total socio-economic system. Income distribution is tied to various economic strata proportionally as those strata own the factors of production. Ownership of the means of production is itself interlinked with the political power structure. It is because of the total socio-economic system that we witness so frequently in developing countries the phenomenon of 'growth without development'. Thus, unless a country has socially progressive programmes and policies, projects can, in fact, widen the existing large income disparities and lead to increasing social tensions.

Another important way in which institutions affect projects is through the whole decision-making process. In bureaucratic systems, the process is painstaking and highly centralised. This relates especially to the financial and budgetary process, where rigid central control over procurement of the simplest goods and services, considerably frustrates those who manage programmes and projects at the field level. The project has to beat the bureaucracy if it must maintain its projected schedules of implementation, and this gives rise to efforts to by-pass existing agencies and create new ones.

Government institutions

The operations of the whole administrative system must be understood. Commonly encountered are situations where there is a well-established balance between the rights of the state or provincial governments vis-à-vis the central or federal government. The authority relationships are sometimes tied to finance and budgetary practices. In Malaysia, for example, land is held by the states and not by the central government. In fact, land constitutes a major source of income for the states through leases, and a variety of land taxes. The central government requiring land for its federally financed projects, must go through the same steps as any individual or private corporations, to obtain the land.

In Indonesia, the provincial governors historically have exercised the right of control over movements of paddy and rice out of their provinces. This limits the development of a national market for rice in the absence of free flows that would normally balance surplus and deficit areas. This certainly did not facilitate the operations of the government's procurement agency, BULOG, as far as the stabilisation policy for rice prices was concerned. Provincial governors also have administrative authority over the staff of the agricultural extension service in the province, who report on technical matters to the centre. This has implications for project preparation.

Within the province or state, there would be further administrative sub-divisions down as far as the village level and the cooperation of the administrations at these lower levels may be required for a successful project. It is generally advisable, where possible, to adjust the project area so as not to straddle too many administrative units.

Sometimes, where there may be many different agencies and departments at the central and the state level concerned with a project, decision-making can be a complex matter involving long negotiations. A common example is where the centre contributes to the financing of a project within a state.

Advantages The main advantage of using the existing government set-up is that this will be the least disruptive of the *status quo*, since it introduces no fundamental changes in the administrative machinery. It, therefore, does not generate vexatious problems of inequality among various branches of the civil service and, thereby, does not set in motion destructive forces aimed at deliberately scuttling the project.

A further important advantage is that the project, through its larger flow of resources, serves as a powerful instrument for strengthening the development role of the department or division. It motivates the staff who prefer the challenge of having a concrete job to do, with prospects of visible results, instead of paper pushing, and this promotes job satisfaction. Management development is encouraged by enabling the staff to offer a larger bundle of goods and services to farmers, instead of just advice, and this equips staff for a larger development role in the future.

Disadvantages Unfortunately, projects do not readily lend themselves to being fitted neatly into departmental/divisional lines. The classic example is in irrigation projects, where the civil headworks and major irrigation structures are executed by the irrigation department. Yet, the benefits to irrigation development are realised only on the farmer's fields, for which the department/division of agriculture must have primary responsibility.

The location of projects within ministries of agriculture might tend to suffer from the fact that the latter are usually politically weak, and thereby, lack the will to promote their projects vigorously in the face of competing claims of more powerful ministries and departments of government. This is generally reflected in their low shares of the government's budget, poorly equipped extension staff and generally unsatisfactory working conditions. Furthermore, the ministry itself suffers from the general bureaucratic malaise of most governmental institutions where speed and timing, so important in agricultural operations, may not be readily appreciated. Staff-recruitment procedures make it virtually impossible for the ministry or department to compete in the open market in order to get the right man for the right job at the right time. Finally, civil service departments and divisions may not be the appropriate place to locate projects whose primary objectives are revenue earning or profit making.

Quasi-public bodies

These are usually created by some enabling law, to undertake specific functions and activities which the government feels are in the public interest. Although falling outside the stream of the country's civil service, and enjoying greater autonomy, both in the conduct of their operations and in their financing, they are expected to implement government's policy in a particular field.

They may take the form of a board, an authority, a corporation or special agency. Organisationally they are attached, however, to some ministry or they may report at even higher levels, e.g. the cabinet or the country's chief executive.

Boards

In Jamaica, there are for example, the Coffee Board, the Coconut Industry Board, the Sugar Industry Board, the Cocoa Board, the Banana Board, the Citrus Industry Board and others. Except for the supervision exercised directly by the Minister of Agriculture, the Boards' functions are so all-embracing with respect to the crop, that the activities and functions of the Ministry of Agriculture are severely curtailed.

The dividing line between boards and corporations (later discussed) is sometimes rather thin and while the functions of the two may overlap, the main identifying feature appears to be that corporations are used when the production aspects are to be emphasised, whereas the board is used to implement the policy of government with respect to a particular product or area, especially where a large trading component may be involved.

The Jamaica Banana Board, for instance, plays a regulatory role in the banana industry. It monitors prices to growers, and is the arm whereby, among others, government grants subsidies for replanting programmes, and takes other measures to control quality and watch the marketing interests of the industry. Since a very large part of Jamaica's farmers are engaged in this industry, there are welfare considerations in addition to economic ones.

Advantages The advantages are that boards have a special role to play, where the regulation of the industry is in the public interest. Their most prominent use appears to be in the marketing field. For example, the Ghana Cocoa Marketing Board provides for orderly marketing, and exercises control over the country's cocoa marketing, since the product is the main source of foreign exchange, and is largely a smallholder operation. Again, with efficient management, boards have a role in stabilising a particular industry, especially where large growers, who can better organise their marketing, are not very much involved. The boards, sometimes through a duty levied on exports, can contribute substantially to the industry's research requirements – an activity that can yield considerable benefits.

Disadvantages The main disadvantages arise from a confusion between or lack of clarity in possibly conflicting policy objectives. Political interference, geared to welfare considerations and vote-catching, can lead to a wastage of resources and provide a continuing cushion of inefficiency in the industry, making it lose its competitive position in export markets. Since normal market forces do not operate, the more efficient growers are penalised for the inefficiency of the others – incentives to increase productivity and quality can be stifled to the detriment of the whole industry. The industry's survival, therefore, depends on a continuous injection of subsidy as grants or low-interest loans.

A proliferation of such boards covering the major products, can thus lead to a situation where government policies contribute to a stifling of initiatives in the agricultural sector. The answer could well lie in a careful choice of the activities that the board should undertake in regard to the particular industry, leaving the non-regulatory aspects, e.g. processing, to the private sector, which would probably be more competent and cost-conscious. However, the whole role of boards and corporations is closely tied to the political direction in which the country wants to move.

Project authorities

An authority is established by a special enabling law when it is decided to vest the implementing agency for a project or programme, with special powers to undertake its task. The authority may be a permanent institution, e.g. a Port Authority, or it may be an *ad hoc* body, with a prescribed lifetime, to accomplish certain defined tasks for which the usual administration machinery was thought inadequate at that particular stage of development.

The purpose of establishing an authority is to ensure coordination among the various government agencies involved, by the concentration and centralisation of authority, so that the project's objectives are realised. In effect, the project establishes its own administration, parallel with those of the existing regional, district and local services. The participating agencies, therefore, yield to the project authority a share of the authority they exert over activities in the project area, which are then administered by the project. An authority, therefore results in the creation of two parallel streams of government. The new authority is usually well stocked with autonomy; with financial resources; with considerable technical skills, often expatriate; and with a clear-cut task to perform over a stated time period.

Advantages Apart from a concentration of skills and resources, the main advantages are that the authority enjoys a shorter chain of command. Decision-making is facilitated and it is easier to schedule activities. The decisions taken at the highest levels of the authority are rapidly passed down and are binding on the project officials lower down the chain of command, at the regional and district levels of the various participating agencies. Since the project's chain of command lies outside of the existing administrative system, officials of the latter have to abide, though reluctantly, with the decisions handed down by the authority.

Again, because the authority has its own resources, concentrated on specific actions within a narrowly prescribed area, it is able to demonstrate results rapidly, in contrast to the existing administration with limited resources and much more broadly-based tasks.

Disadvantages The largest single disadvantage of the fixed-term authority is that it is an organisational intrusion. It stands on its own, without being grafted on to the existing system. Therefore, it is competitive and not supportive, and because of this, it generates frictions and jealousies. Frictions arise for several reasons, the most important being the privileged status enjoyed by the personnel of the authority.

Because the authority is concerned with the physical achievements of the project, the objectives of the management are to get visible results quickly. Institution-building thus takes on very little or no priority at all in management thinking. At the end of the project's development period, assets have been created but their maintenance and operation must revert to the existing administrative structure, which, having felt left out initially, lacks the will, enthusiasm or even skills to sustain the project in its post-development period, when real benefits are to be derived.

Of great importance is the fact that the existing administrative structure does not get experience in development projects. This requires very close study of the authority as an instrument of development growth. It does not set into motion growth-sustaining forces, so important to the development process.

A further problem with the authority approach is that it may accord priorities to certain project activities within the project area, which do not harmonise with the regional priorities of the various participating agencies, and may drain resources (finance and staff) away from other priorities in the country.

In summary, therefore, project planners have much to learn about how to coordinate the involvement of several disciplines and ministries. Until management students and public administration experts can establish some working guidelines, the authority approach will continue to be a divisive factor in the agricultural-development process, particularly where the country feels that this approach is being enforced as a condition to receiving external aid.

Corporate entities

This group includes corporations and companies. There may be differences between countries as to the various functions that such corporate entities may be empowered to undertake. However, all corporate entities are there to introduce a business approach to problems entailing clearly commercial objectives which fall outside of the usual stream of government activity. A corporation may play a facilitating role by approving projects and making funds available to the private sector for those activities that the public sector wishes to promote. Also, it might engage directly in the production process. The objective of the corporation is to bring not only financial resources but managerial skills and improved technology to bear on a particular problem. Thus, in agricultural projects, the purpose may be to bring the benefits of the estate approach to small-farmer production.

Advantages One advantage of the corporate organisation is that it is best suited to the exploitation of existing resources. Examples are: fisheries development, processing and marketing. In forestry, it is well suited to logging operations, and to plantation establishment with a long gestation period. A corporation is also well suited to projects, e.g. a pulp and paper mill, where large capital investment is required, and domestic private or foreign enterprises would like to get government to share the risk. They are also useful when a government wishes to promote rapidly the growth of a particular product or agricultural industry that may be relatively new to the country, e.g. cotton – that is, where it is appropriate for the public sector to undertake the risks and costs of innovation, so as to induce private investment later.

Disadvantages The main disadvantages are that in the absence of sound management, corporations could become a drain on public resources. In one Asian country, a fisheries corporation with US$50 million of assets, had to be liquidated and only US$6 million were realised. Sometimes with too many corporations exerting considerable control over activities and assets, they can function almost as a government within a government. The autonomy that they enjoy creates tensions within the usual administrative stream. Since corporations are usually profit-oriented, and since their management may not be rewarded on the basis of profit performance, there is a lack of incentive to work as hard at gaining profits as would be the case in a privately operated firm.

Land-tenure systems

These are of crucial importance to project work. This feature of the institutional system is the one that most affects the distribution of the project's benefits by beneficiary classes. The land-ownership patterns and landlord–tenant relationships powerfully influence production, employment opportunities and income distribution. Apart from this, farm-size distribution is an important feature in the intensity of utilisation of farm land, and in motivating farmers to undertake innovations.

Project planners are usually happy when they find the project area largely owner-oriented, because it simplifies the task of implementation of project actions. This affords a neat solution to the problems of the tenant farmer and the landless. This landless group now comprises about a third of the farm population in the developing world and is treated as the 'silent majority'. This group may well begin to pose serious problems in the coming decades as population growth increases their numbers.

The small farmer also, until very recent times, has been by-passed by commercial lending institutions within the countries, and international development institutions and donors in their operations in these countries. He has been left largely to fend for himself against the landlord where he does not own his land, against his creditors, and against middlemen operators in selling his product.

Because of the impact of tenure on the project's design, data are required for the project area, about the distribution by size of various farm classes; the land-ownership classes including freehold tenancy and rental conditions; and the existence of communally held lands. Where no recent census data exist, sample surveys may have to be undertaken. For a seeds project in India, the Agricultural University of Uttar Pradesh undertook such a survey for the project area with the findings shown in Table 8.1. This sample represented some 50 per cent of the possible project participants and confirmed the availability of large farm-sizes in the project area – a situation quite favourable for the production of seeds. For a seeds project, intensive supervision must be provided to ensure product quality, and project implementation is favoured here by the presence of a number of large farm-holdings in the sample.

TABLE 8.1 India: Uttar Pradesh – Farm size distribution

Size range (ha)	**Number of holdings**	per cent	**Total hectarage**	per cent	**Average hectarage**
Less than 20	65	31·7	0·841	5·5	13
21–40	57	27·8	1·916	12·6	34
41–100	46	22·4	3·077	20·3	67
101–200	27	13·2	4·120	27·2	153
over 200	10	4·9	5·209	34·4	521
Total	205	100·0	15·163	100·0	74

Large farm-sizes can have both positive and negative implications for a project. Where mechanisation is planned as a project component, large farming units are helpful. They may also be indispensable for beef-raising enterprises, and can be especially important where an industrial end-use is contemplated, for example, in palm oil production. On the other hand, large farm-sizes could have negative aspects in an irrigation project, where the larger operators may not, for various reasons, wish to bring their whole farm area under irrigation, and therefore contribute to the under-utilisation of the water resources.

Finally, owners of large farms may show significant differences from the owners of small farms in such features as, attitudes towards risk-taking in introduction of innovations, managerial and farming skills and access to lines of credit, all of which could exert an important bearing on the project's success. Moreover, the owner of a large farm is probably much more market-oriented in his production planning than the owner of a small farm and, therefore, more responsive to general price movements and demands facing the commodities he produces.

Where the project analyst is concerned with the income-distribution aspects of the project, as in rural-development projects, it is important to know within reasonable limits, the distribution of landholdings by size in order to identify the target group of the project's beneficiaries. The tenancy terms and conditions may be equally or even more important than the farm size. The more important elements cover the length of the land rental period, the levels of cash payments, the types of crop sharing arrangements that prevail, and the general landlord-tenant relationships.

The project planner must, therefore, have a reasonable knowledge of the tenure conditions in the project area, so that he can not only identify the potential beneficiaries, but also assess the incentive/disincentive impact of the various project actions and the possible production responses. Sometimes farmers with short-term rentals would not apply fertilisers because they feared that the fertilisers would still remain in the soil to the landlord's benefit, after the season's crop had been harvested. The tenure system can further operate as a major constraint in the requirements concerning credit-worthiness enforced by credit institutions, thus posing problems for credit projects geared to small farmers.

Banking and credit

The following brief treatment cannot possibly do justice to the importance of agricultural credit in the process of agricultural development. We are concerned here, less with the overall policy issues, than with the impact of the existing credit institutions and structure on the operations of a particular project.

Some broad generalisations are called for, however. First, agricultural credit is employed in many countries as an arm of broader agricultural policies. It is not uncommon to find that interest rates to agriculture especially small farmers, fall below the opportunity cost of capital in the country. While the apparent objective may be to stimulate agricultural production, the real result is one of subsidising the group, in effect some muted form of income distribution. Generally, such programmes tend to be small and have little impact on the total agricultural situation since non-institutional credit continues to dominate the picture. Most governments, in their credit programmes aim to by-pass or dislodge the established non-institutional credit systems, about which so little is known: for example, how are interest rates determined? A new objective might well involve new approaches that incorporate the established systems.

Second, the major problems in credit lie in the small-farm sector. Owners of larger farms, even under imperfect conditions, always have access to loan funds. In Jamaica, for example, the bulk of the owners of larger farms finance their long-term plantation development with short-term annual borrowings, employing overdraft facilities from the commercial banks.

Lending to the owners of small farms is fraught with difficulties. These include:

1 high administrative costs of lending per farmer;
2 lack of knowledge by the farmer of how to employ credit funds for productive purposes;
3 the absence of adequate technical guidance from the extension services;
4 the problem of adequate collateral and other forms of security;
5 the generally low recovery rate leading to a loss of capital and of liquidity of the lending institution;
6 the uncertainties of yield and production arising from weather, floods, etc.;
7 uncertainties arising from market and product price fluctuations.

In some countries, attempts have been made to overcome the recovery problem by a system of collective responsibility of the loan beneficiaries. Finally, agricultural credit in many countries is sometimes viewed by both farmers and politicians as a political weapon rather than as a stimulant in the development process, and the owners of larger farms are sometimes the first to default on their loan repayments.

It must further be recognised that production credit for agriculture is only one facet of the farmer's total credit requirements, and his consumption credit needs for himself and his family may considerably outweigh the former in importance. This binds him, therefore, to the village grocer or suppliers and the old adage of 'I owe my soul to the company store' remains very pertinent. In such a system, where the farming family's requirements are advanced in cash and in kind and repaid in produce at harvest time, the true costs of credit extended can hardly be measured with any precision. In fact, a paternalistic bond may prevail between the credit grantor and the grantee, arising from an 'insurance element'.

During project formulation, the project planner is usually concerned with a credit element, from some source to the target group. He must differentiate between a straight agricultural credit project and other agricultural projects with a credit component. For example, irrigation, livestock, general agriculture and rural development projects, could also have a sizeable credit element. A major problem exists where farmers in the project area may already be heavily indebted to other creditors and the issue of refinancing may well arise. Commercial banks, in most countries generally accord low priority to agricultural lending, except in countries where an established estate agriculture exists. A major reason for this is that they could well view agriculture as a high-risk enterprise, and that in any case the yields from investment in other sectors, e.g. construction, manufacturing and trading operations are generally higher. In some Latin American countries, financing of cattle operations using the stock as collateral might prove elusive since cattle can walk and be slaughtered.

Farmer organisations and cooperatives

So much has been written on the subject of cooperatives that we do not propose to elaborate on this. Success stories of multi-purpose cooperatives are extremely limited. These cooperatives undertake a wide range of activities covering production, marketing, input supply and distribution, and operate their own extension services. Apart from this, the experience of cooperatives, largely single-purpose, in most developing countries has been disappointing on the whole.

Several reasons account for this. The cooperative may have been imposed from the top and the services offered or its objectives may not have obtained the required support at the grass-roots level. In some countries the government may actively discourage associations or groupings of farmers since they may constitute possible foci for political infiltration. At the working level, however, a major reason for failure has been the deficient quality of management, particularly where the cooperative undertakes agro-business type operations. An unskilled, poorly paid management with a lack of business experience, is quite common, leading to financial losses and disappointment of the membership. Deliberate institution-building activities by aid-donors and lending agencies can do much to correct this situation.

Where such associations and cooperatives exist in the project's impact area, the project analyst is well advised to explore the possible role that they could play, after making necessary provisions for strengthening them. A basic principle in project work should be to ensure that a cooperative does, in fact, meet a basic need of the bulk of the participants. A good example is water-management in an irrigation tract. The functions should be kept as simple as possible. Despite repeated failures, cooperatives are still widely held to be the answer to many rural problems. Project planners should generally be cautious in this regard, and critically examine relevant experience in the country and the project area.

9 Organisation and management features

It is generally conceded that the organisation and management arrangements for a project are the most difficult to formulate and possibly the largest single traceable source of problems during the project's implementation phase. In fact, we would surmise that these problems account largely for the wide discrepancies between loan funds committed and actual disbursements, that constantly face both the lending institutions and the borrowing countries.

Despite this, there has been so far no systematic attempt to evaluate the abundant experience that now exists, to identify the basic issues and problems, and to explore new approaches to the design of management systems suited to the various environments. This is a fertile field where students of business management, public administration experts and practically oriented sociologists could make a major contribution. Before proceeding with the identification of the problems and examining some standard approaches, an understanding of the terminology is important.

Organisation

The term organisation essentially refers to the overall design and structure of the body or entity that would undertake the task of project execution. For simplicity we can compare this to the architect's design of a building which, beyond being functional, must meet the ecological requirements of the environment and must conform with the building regulations of the place in which it is sited. In the same way, the proposed organisation must have the capacity to carry out the assignments given to it, and also be attuned to the institutional and legal framework of the country. It must also take cognizance of the political and social climate within the country.

Thus the organisational structure must reflect the objectives of the project:

> The organization planner begins by knowing where his company wants to go; organization is a plan for getting it there. The objectives determine the structure needed.[1]

In effect, what the organisation structure does is to establish relationships and the legal framework within which the management of the entity will have to work to achieve the project's objectives.

Selecting an organisation

In defining the organisational arrangements for a project a number of options may be explored. These may range from the use of an existing institution, strengthened if necessary, to the creation of an entirely new organisation. Organisational arrangements must depend upon an examination of many factors, including the following:

1 The existence of institutions already empowered to carry out the proposed functions.
2 The capacity and efficiency of these institutions, their success in assuming these functions, and the priority which they would be able to attach to the project activities.
3 The sensitivity of components of the project to local conditions.
4 The urgency of implementing the project.
5 The extent to which an independence from governmental bureaucracy is a prerequisite for the successful implementation of the project.

While planners are frequently tempted to propose the creation of new organisations to implement their projects on the grounds that the existing ones are hopelessly inefficient, this may be a short-sighted approach. Too often new institutions:

1 Duplicate the functions of existing institutions, which nevertheless continue to exist even after the birth of the new one.
2 Take longer than expected to become effective, because of the need for enabling legislation.
3 Surround themselves with special conditions (especially those relating to salaries) which may enable them to recruit well-qualified staff but which create distortions in the civil service structure of the country in question and set off a chain reaction of claims for preferential treatment.
4 Set precedents for the creation of other similar institutions all with claims of access higher than normal points of decision-making, until the decision-makers themselves become bogged down by the proliferation of 'authorities' and 'commissions' to which they are linked.
5 Become obsolete, but self-perpetuating, once the project has been implemented.

Nevertheless, the execution of certain tasks does require the creation of an entirely new agency, but this of course should be favoured only when there is no existing institution with appropriate functions, or where those endorsed with the functions are unable to cope.

The form of an appropriate organisational structure is usually governed by a wide range of factors. The more important are: the need to conform or not to civil service regulations covering, for example, salary and wage levels and promotions; the capacity to contract debt obligations, both domestic and foreign; the use of cost-accounting practices; special concessions and immunities with respect to income taxes, export and import taxes; entitlement to subsidies through the government's budgetary process, and other concessionary finance (e.g. subsidised interest rates or special foreign exchange privileges); powers to engage contractors and consultants; powers to acquire and dispose of assets; powers to sue and to be sued, and powers to operate relatively independent of the country's treasury. Availability of personnel must be considered and opportunities to recruit capable people must be explored.

Internal organisation

Apart from the choice of an organisational structure, there is the separate question of the internal structure – that is, the departmental, divisional or unit structure. This, in turn, should derive from a clear identification of the various activities to be performed and their grouping together usually on a basis of similarity. For example, in a seeds project, we may identify broad groups of activities as follows: policy formulation, regulation (inspection and control) and commercial activities such as seed production, purchase, processing and marketing. In an irrigation project, the main activities may involve design and construction of civil works, land development, water management, farm mechanisation, delivery of inputs, research, training of farmers and credit administration.

In determining the most appropriate organisational structure, a number of basic management principles must be applied. Some of the more important ones are as follows:[2]

1 There must be clear lines of authority running from the top to the bottom of the organisation. That is, the chain of command should be clear.
2 The responsibility and authority of each supervisor should be clearly defined in writing.
3 Responsibility should always be coupled with corresponding authority.
4 Authority should be delegated as far down the line as possible. This implies that the decision-making power should be placed as near as possible to the scene of action. (This principle is frequently violated in projects.)
5 The number of levels of authority should be kept at a minimum.
6 The work of every person in the organisation should be confined as far as possible to the performance of a single leading function. This is the principle of specialisation.
7 Line functions should be separated from staff functions whenever possible. The line functions are those which accomplish the main objectives of the enterprise, while staff play an auxiliary or advisory role.
8 The organisation should be flexible, so that it can be adjusted to changing conditions.
9 The organisation should be kept as simple as possible.

Apart from the above principles, one other basic principle needs emphasis. There appears to be an inherent tendency among most project planners to neglect the principle that an organisation should not be designed around personalities. Too often the availability of a particularly capable individual tends to impose a bias in the decision-making process as to where to locate a project in a particular context. A major reason for this is the severe shortage of staff, in the developing countries, capable of implementing projects in agriculture.

The country's political and administrative system is a major consideration. One of the issues most frequently encountered in the design of projects, relates to the extent to which the implementing organisation must be decentralised to be most effective. Frequently it is difficult to reconcile the advantages of securing the involvement of the decision-makers with the need to delegate authority to staff near the site of action, and to make a project's management sensitive to the views of the beneficiary population. Certain project functions may lend themselves to very strong decentralisation, while others require centralisation. Decentralisation refers to the extent of authority delegated down the chain of command for taking decisions. It is not the number of decisions that indicates the degree of decentralisation but rather the importance of the decision, e.g. a decision that binds the resources of the project. Decentralisation is most required in countries with poor communications, as, for example, Nepal in a highly mountainous country with a poor road network and poor telecommunications system. It is also important in countries where decision-making is apt to get bogged down by archaic bureaucratic systems, where the smallest decisions may conventionally be taken at top government levels. The area of procurement of goods and services is a fairly common example.

The organisational structure may, therefore, aim at centralising broad policy decision-making as it relates, for example, to the annual programme to be undertaken; the marketing strategy for the project's output; procurement of major pieces of equipment; negotiations with contractors and consultants and setting a recruitment policy. The mechanics of the project operations, for example, how and when to plant particular crops, or whether to grant a credit loan should be strongly delegated.

Beneficiary participation

Although it is widely acknowledged that the involvement of the potential beneficiaries of a project in its planning and operation is a prerequisite for success, the design of organisational arrangements to secure such involvement is particularly difficult. In planning the project it is necessary to identify the structure of the social authority operating in the area and to explore with their representatives – the leaders of cooperatives, irrigation water-management groups or traditional credit societies – various possibilities of involving them in the planning and management of the project.

The method of securing such involvement will vary. In the Yemen Arab Republic, a democratically elected Local Development Association movement has emerged as a means of mobilising people to participate in executing rural infrastructural works. The approach here has been to form a committee at provincial level in which the Association's representatives meet, under the chairmanship of the borrower, together with the representatives of a Project Authority. The latter can provide them with the engineering expertise they need to complement their own political and financial strength.

Management

It is the task of the management to implement the project's objectives within the framework of the organisational structure provided for the project. The organisational framework may be an existing government agency or agencies, an agricultural credit bank, a development corporation, or a specially created project authority. A good management can make a deficient organisation structure work but the reverse does not hold.

There are four basic considerations in designing a management system for a project.

Definition of the objective A clear definition of the functions and activities required to achieve the project's objective is the first consideration. In a project for irrigation and agricultural development, we could envisage the following major activities; the on-farm development works such as internal drainage and land-levelling; the provision of agricultural support services such as extension services, agricultural credit, farm input supplies, farmer training and arrangements for product marketing.

In such a project, where the larger bulk of the project's investment would entail the construction of civil works, it is clear that the country's irrigation department, or authority would have to play a leading role in project implementation. However, the involved agency may well view its task as ending with the distribution system up to the tertiary canal level. The project's full production potential would not be achieved without the host of other activities indicated above.

Allocation of responsibility To allocate and assign responsibilities to various agencies or associations for the various project activities and to spell out clearly the duties including the undertakings that the agency would be required to implement under the project. An illustrative example is a rural development project in the Philippines, as follows.

The National Irrigation Administration (NIA) would be responsible for the national and communal irrigation system and it would, jointly with the Bureau of Forest Development, be responsible for the watershed protection programme. The NIA would strengthen its staff to execute the project (located in Mindoro) and would provide engineering, design and planning support from its central staff in Manila. There would be five water-management divisions, with a supervisor responsible for each division, which would be divided into blocks of 500 ha, each controlled by a graduate water-management technologist. Each technologist would control five ditch tenders, who would each supervise two 50-ha irrigation units.

The Department of Public Highways would be responsible for national highway improvement and maintenance. It would prepare detailed designs and final engineering specifications and tender documents, award contracts and supervise construction. The provincial governments would provide adequate staff, facilities and funds to maintain and repair the provincial roads for which they are responsible.

The Bureau of Agricultural Extension, Bureau of Plant Industry and Department of Agrarian Reform all provide extension services in Mindoro. The extension staff would be increased to reach a target of one technician for every 300 farmers in the two provinces. They would be supplemented by a further 24 water-management technologists.

In addition the project envisaged the formation of pre-cooperative farmers' organisations based at the village level. These groupings of 20–25 farmers would serve as channels for the inflow of technical services, credit and inputs and as assembling points for produce for marketing. Extension agents would visit each organisation about once every two weeks while each organisation would appoint a progressive farmer to assist the technicians and maintain contact with the farmers between visits.

Coordination of activities To formulate a mechanism for coordination of the activities of participating agencies. This is especially critical in rural development projects which are characterised by a wide range of diverse activities occurring simultaneously. For instance, in the project discussed above, several agencies besides those described above were involved.

The project proposed a two-tier arrangement for coordination at the national and provincial levels (Figure 9.1). At the national level, the main line departments and agencies comprised the National Irrigation Administration (NIA), the Departments of Public Highways (DPH), Public Works (DPW), Agriculture (DA), Public Health (DH), Local Government and Community Development (DLGCD), Natural Resources (DNR) and the Presidential Assistant on National Minorities (PANAMIN). A Project Director, reporting directly to the Cabinet Coordinating Committee for Rural Development (CCC) located within the National Economic and Development Authority (NEDA) would be responsible for coordinating the activities of the involved agencies and for implementing the project at the national level. The Project Director, assisted by a small technical and management staff, would be responsible for approving detailed annual budgets and work programmes for project activities, prepared by the participating departments and agencies. The coordinating mechanism for this would be a Steering Committee which he would chair, and which would comprise senior officers from each department and agency involved in the project and the two provincial Governors. Each department and agency would prepare an annual budget and work programme for its part of the project for approval by the Steering Committee. It is important to note that this Committee enjoys executive authority and does not merely play an advisory role since its major task is to ensure and commit financial resources for the project's implementation.

At the provincial level, the coordinating mechanism would be the Provincial Development Council chaired by the Governor in each province (Figure 9.1) and comprising the Provincial heads of line agencies. As conceived, the senior line officer of each line department and agency in the province would be the project officer for his organisation and would be a member of the Council. The Council's main task would be to resolve problems of inter-agency coordination at the local level, and is in effect the provincial counterpart of the national level Steering Committee. The provincial Governors would have direct access to the Project Director as the occasion arose.

Staffing the project Staffing is defined as:

> The executive function that encompasses the recruiting, selection, training, promotion and retirement of subordinates. The attention the executive gives to the development of managers is indeed a present cost of providing for the future welfare of the enterprise.[3]

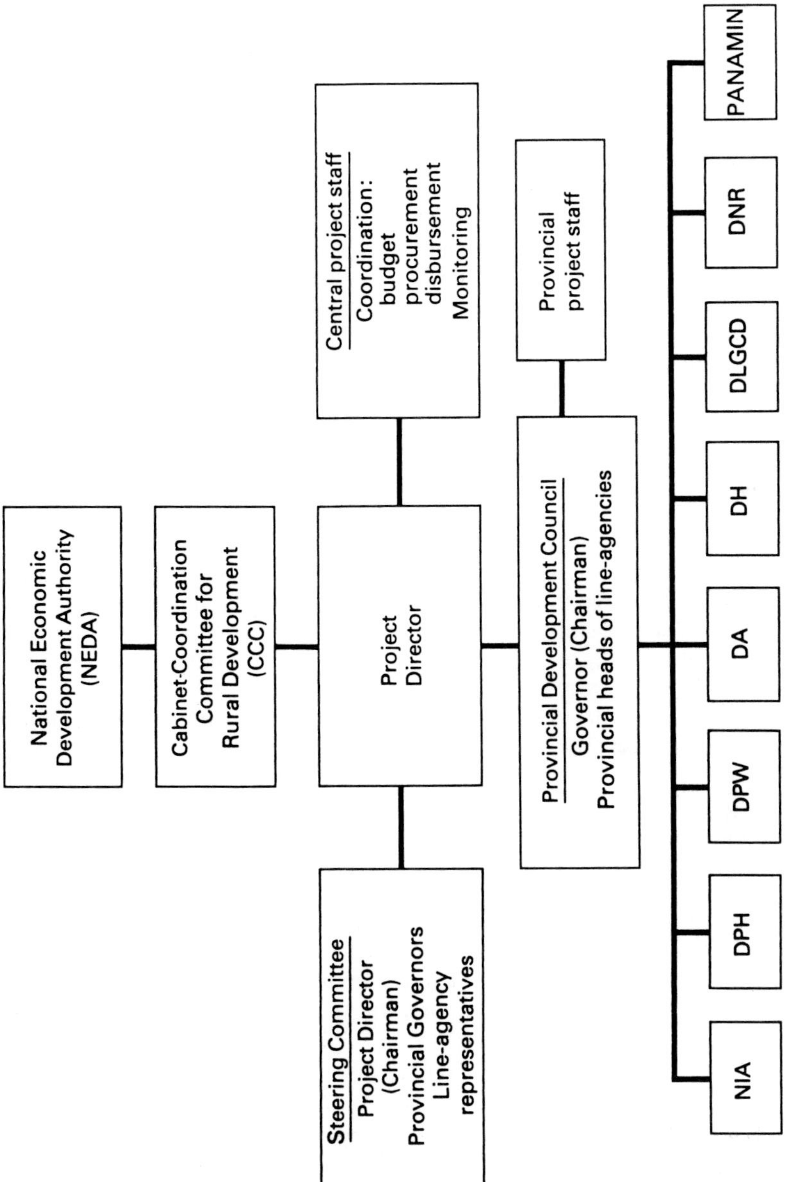

Figure 9.1 Philippines: Rural development project – Project organisation

Many projects attempt to overcome the problem of staff shortage by inclusion of varying amounts of expatriate technical assistance during the project implementation phase. However, a consultant brought in under contract to do a job in a country, receives payment on the basis of the fulfilment of the job specifications in the contract. While it is not infrequent to include the training of local nationals in the consultant's terms of reference, there is inadequate control and sometimes not even the slightest check is made to ensure that this has been done. Thus the physical infrastructure always moves ahead of the development of local managerial skills required to get the new assets truly productive.

Nowhere is this situation more apparent than in irrigation projects financed by international lending institutions and donor countries in many parts of the world. Since there are an extremely limited number of international consulting firms that can offer a country expert service in irrigation management, the developing countries are left to sort out the problems themselves. Their only alternative is to depend on individual consultants selected on an *ad hoc* basis, who may lack the technical back-up so vital to the performance of an expert in a country.

The staffing problem in projects is not simply one of finding a few managers at the uppermost levels, but equally important is the manning of the organisation structure with the skilled and semi-skilled workers required. This involves, for example, trained extension agents, land surveyors, tractor drivers, accountants, sawmill operators, etc. In Liberia, a major constraint to the Liberianisation of forestry and forest industries is the acute shortage not only of Liberian entrepreneurs who know the business but also of skilled workers for operating the heavy and expensive equipment for road construction, for logging and for extraction. The gap has to be filled largely by reliance on Ivorian operators.

The staffing problem that developing countries face in implementing their agricultural projects arises from many sources. First, there is the manpower shortage both in quantitative and qualitative terms. Second, there is the absence of motivation for qualified staff to work for protracted periods of time in remote agricultural areas. The Civil Service regulations of the country may not permit the granting of special allowances as an inducement to staff to live with their families in areas where living conditions may be unattractive. In countries such as Malaysia, with a strong private agricultural sector, it is not unusual to find that as soon as a trained cadre of staff is built up in the public sector, it is quickly attracted away to the private sector with the latter's more lucrative job assignments.

Choice and evaluation of implementing agency

A few basic types of organisational structures, usually employed for the management of financed projects, are discussed below largely in terms of performance evaluation.

In the case of government departments, project implementation lies within the mainstream of the existing administrative structure. This choice is especially appropriate where the bulk of the project's activities involve one basic discipline. Thus, a livestock project might be located within the livestock department or division of a Ministry of Agriculture; an irrigation project within the irrigation department; and an agricultural credit project within an agricultural credit bank or fund. In such cases, the existing staff structure may frequently be used without any significant changes, or a special project unit may be created within the existing structure, with specific responsibility for the project's execution. The actual works could be executed on force account or by contractors and consultants.

In any event, the project planner will have to appraise the performance of the division and the individuals who would occupy executive positions in the project's management. The appraisal is relatively easy where the project actions are designed to support an already on-going departmental activity. It involves the assessment of actual performance against targets, and determining reasons for shortfalls. For example, budgeted funds may not have arrived in time, or considerable delays may have been entailed in procurement of equipment. The judgement must be based on an assessment as to whether the resources under the division's or department's command were efficiently employed and whether the planning objectives were realised.

Where an existing corporation or company is being considered as the implementing agency for the project, then it would have to be appraised as follows: what has been the general performance to date, and what are the factors that determine performance? For example, does it operate in a sheltered environment or is it competing in the open market with private sector firms with similar interests? How does it measure when appraised against the objectives for which it was established, with regard, say, to its impact on agricultural production, on foreign exchange earnings, on employment, on price stabilisation, on increased farmer income; and at what cost has its achievement been obtained?

Is the financial structure sound? What is its debt equity ratio, its return on assets and sales, and its general operating performance? Is the corporation authorised to contract foreign debt, or can it issue bonds in the local market? Does it represent a drain on government's resources in the form of continued subsidies? What is its general standing in the mind of the public? What is the quality of its management? Does it have an effective board of directors capable of independent decision-making or is it prone to serious political interference that impedes its decision-making processes? Is there a balanced representation of the various interest groups in the management of the corporation or company? Are reports issued on time and are there satisfactory arrangements for audits?

While a similar type of analysis would be used for boards, there would be more concern for issues relating to the success that the board has had in interpreting and implementing public policy on matters within its competence. If, on the other hand, no existing entity exists, and it is necessary to create a new one for the project, then the planner has to establish the legal basis for doing so. This has to be done jointly with the country's legal experts, so as to ensure that the charter for the proposed new entity conforms with the country's laws and will equip the entity with the required powers to achieve the project's objectives.

The question of who manages the project becomes fundamental. Faces and names come into play as jobs have to be matched with potential candidates and the hierarchy must be reckoned with. Assuming for example that the livestock division is headed by a veterinarian, not uncommon, with no feeling for or application of broader livestock production issues, the project's success might be thwarted if he were selected as the manager of a project with a main thrust of pasture development, water supplies, abattoirs, etc. On the other hand, he may perform admirably if the project was highly research-oriented or had a strong disease-control component.

The role of committees

Committees do, generally, play an important role in the planning and execution of agricultural and rural-development projects. They are useful vehicles for ensuring representation of interested groups; for purposes of coordination; for a transmission of information; for motivation through participation, and ensuring adequacy and maintenance of the project's technical standards. As far as project work goes, the committee has a definite role to play in the airing and discussion of broad policy issues, so that the government's varied considerations (including political) can be brought into focus in weighing decisions affecting the project. They can also play a coordinating role between the project's organisational structure and the established administrative and social structure at the district, village and project area levels.

Committees, however, have severe limitations. For example, it has been suggested that the reason a camel looks the way it does, is that it was put together by a committee. Probably the first question to resolve is exactly what role the committee should play. If it has a significant role to play, then it should be formally built in to the organisational structure. There may sometimes be place for a committee with a purely informal role. Assuming the presence of a formal committee, the question arises whether it should enjoy line authority or merely staff authority with an advisory role,[4] and at what level within the organisational structure it should be located.

There is a potential conflict here to be recognised. The higher up the decision-making level that this is located, the higher the level of representation must be from the different ministries and departments. Unfortunately, a major problem facing committees in developing countries is that the nominated representative sends alternatives, who are invariably in no position of authority to commit their respective organisations. The reality of the situation is that the same people are requested to serve on dozens of committees, and these are the ones whom every visiting mission, governmental and non-governmental, wants to see to get a decision taken, or at least to state their respective cases. It is thus the time constraint that makes the committee system break down. This refers even more so to officials in the ministries of finance and/or planning, where major decisions affecting the project's welfare are usually concentrated.

References

1 Max D. Richards and William A. Nielander, *Readings in Management,* South-West Publishing Co., Cincinnati, Ohio, 1958, p. 536.

2 *Ibid.* pp. 539–47.

3 Harold Koontz and Cyril O'Donnell, *Principles of Management*, McGraw Hill Book Co., Inc., 1959, pp. 305–306.

4 *Ibid.* pp. 224–52, for a good treatment of committees.

10 Financial and commercial aspects

The main focus of this chapter is on practical issues involved during project preparation, but several examples are offered to illustrate the methodology used in the financial analysis of project investments.

Financial aspects

Purposes of financial analysis

Financial analysis attempts to measure the financial viability of the project, and is a necessary complement to the economic analysis in decisions to undertake the project. Since a project may have several different beneficiaries and participating agencies, and since it may impinge differently on each one, its financial impact should be assessed separately for each. We must, therefore, differentiate between analysis at the level of the ultimate beneficiary – frequently the farmer but possibly a company or agro-industrial enterprise – and that of intermediate institutions, such as agricultural banks or development cooperatives; and at that of the central government.

The main objectives of the analysis at the level of the farmer or entrepreneur include the following.

Assessing incentive aspects of the project and setting farm-income targets Will the assets created or acquired by the project, result in sufficient additional revenue after debt servicing, so as to induce the investor to undertake the activities envisaged under the project? This is one element in assessing the incentives to farmers to accept change. Are the incentives offered by an irrigation project adequate to induce the farmer to undertake double cropping, which may, in fact, be the main economic justification for the project?

Assessing viability at various stages of the project This is particularly important in integrated operations where the product produced at the farm level becomes the raw material input into a processing component. Assume, for example, a food-processing project including components for on-farm production of tomatoes, and processing. In such a case the financial analysis should answer the question whether the tomato growers can deliver tomatoes to the factory at a price that makes it viable for the plant to operate.

Assessing borrowers' repayment capacity and determining lending conditions The financial analysis should guide the credit institution on the borrower's capacity to meet amortisation and interest payments on any debts to be incurred. Projections of income and expenditure may be used to indicate the most appropriate scheduling for loan repayments, including grace periods, should these be necessary. Debt servicing may be particularly heavy, for example, in land-settlement projects, where the settler farmers have to purchase the land, while also amortising their development loans. The remaining cash balance should be such that the farmer and his family are induced to bring the farm into full production rapidly, rather than seek alternative employment opportunities.

Determining the contribution of beneficiaries to project costs Once the total project costs have been determined, it is necessary to decide the contribution that beneficiaries should be asked to make and what level of taxes or levies should most appropriately be imposed. Establishing measures for recovering public investments applies particularly to capital intensive projects, e.g. irrigation and land settlement.

In assessing the impact of the project on the budgets of government institutions, the more important objectives of the financial analysis include the following.

Formulating the project's financing plan Every project has a financing plan – worked out between the aid-agency and the country – which shows the sources of funds and their application. During the preparation of a project, proposals may be developed for a financing plan, which would indicate the respective contributions of the lending agency, the government and beneficiaries to the project investment programme. Unless, however, the financing agency makes a clear commitment, at an early stage of project preparation, to advance a certain size of loan for the project, it may be better to postpone the preparation of the financing plan until appraisal time. The plan, whenever it is prepared, should be presented in the form of tables indicating, for example, the government's contribution to the project as grants or equity and loan funds, credit funds from external sources, including suppliers' credits; internal market borrowings by the corporation or authority; and contributions from any other sources, e.g. disposal of existing assets.

The end-use of funds should be shown by broad groups, such as machinery and equipment purchases, farm credit, buildings construction, etc. Equally important is the proposed phasing of expenditures, that is, the year-by-year expenditures over the disbursement period.

Assessing the impact on the government's budget The project may affect the government's budget both on the expenditure and incomes sides. The expenditure side may include such items as grant or equity contributions on which no repayment of capital is expected; loans extended at rates below the prevailing market interest rate; subsidies in various forms; costs of services (e.g. extension) absorbed by the government; infrastructure requirements such as electrification, rail connection, roads for which the project would be the main justification but the costs of which are not charged to the project.

The income effects may arise from several sources: increased land and property taxes stemming from the project's investments; export taxes on agricultural products; income taxes on profits arising from the corporation which may revert to the government treasury; and also indirect benefits that can be identified and assessed. Finally, the evaluation must cover the impact of the project's operating costs on the budget of the government departments concerned.

Main concepts of measurement

Gittinger[1] cites three main tools that financial analysts currently use in assessing the probable impact of a project on incomes, namely the benefit–cost ratio, the net present worth, and the internal rate of return; the last one is most widely employed in project work. All three measures involve the application of discounting techniques to forecasts of income and expenditure with the aim of expressing the 'time-value of money'.

The benefit–cost ratio

The benefit–cost ratio is derived by dividing the discounted stream of benefits by the discounted stream of costs as follows:

$$\text{benefit–cost ratio} = \frac{\text{present worth of benefits}}{\text{present worth of costs}}$$

The discount rate usually chosen is one that reflects the 'opportunity cost of capital' in the country. Where the ratio exceeds 1·0 then the project passes minimum acceptable standards. This measure has been widely used in the USA for evaluating water-resources projects.

The net present worth

The net present worth is the present worth of the net annual balance of the cash-flow stream. Differences between the total project's costs and gross benefits in each year of the project's life are netted out and discounted using an interest rate that reflects again the opportunity cost of capital. It is the difficulty of choosing an appropriate discount rate that poses problems for use of this and the previous approach. As long as the present worth is positive the project could qualify to be undertaken. A major disadvantage is that it cannot be used for choosing between alternative projects.

The internal rate of return

The distinction must be made between the internal rate of return on total investments at the level of the enterprise, subject to analysis, and the rate of return on the entrepreneur's own equity investment. Both are relevant to decision-making during the preparation of a project.

The financial rate of return differs from the economic rate of return (see Chapter II) principally because it is normally derived from estimates using actual or projected market prices, rather than estimates of the 'economic worth' of inputs and products. Thus, if a farmer receives fertilisers at subsidised prices, this price is used in the financial analysis. Similarly, on the output side, the prices used for the product the farmer sells are the prevailing farm-gate prices, adjusted, where necessary, to take account of seasonal price fluctuations.

When evaluating the impact of proposed investments on the financial status of a commercial enterprise, it may be useful to extend the analysis to include forecasts of profits and losses, on an annual basis, rather than merely to assess the rates of return on investment.

The farm model

Financial analysis in agricultural project preparation is carried out most frequently at the level of the farm. The detailed consideration given below to farm models serves to illustrate principles of financial analysis which are also applicable with only slight modification at the level of a company, intermediate financing agency or government institution.

The farm model is a predictive tool for forecasting financial performance at the farm level. The model purports to show physical input/output relationships over time, and to convert these into financial terms. In some cases, an aggregation of farm models, with appropriate adjustments for prices, may serve as a basis for assessing the economic benefits of the project.

While the formulation of the farm model includes a large subjective element, the element of subjectivity reduces substantially with increasing knowledge of the project's parameters. The farm model is the crux of the interaction between the technical, institutional, organisational, managerial and other aspects of the project. We attempt to assimilate all this to derive answers to a few basic questions on yields, costs, prices and rate of adoption of changes proposed by the project. At this point, individual disciplines come strongly into play – those of the agronomist, the livestock expert, irrigation engineer and farm-management expert.

The aim of the farm model is not to define a hypothetically ideal situation, but rather to forecast what can be realistically expected to happen in the light of all the constraints facing the project. A farm plan is, therefore, not realistic if it attempts to predict yields simply on the basis of a recommended package of inputs for which evidence may well exist that the farmer will only partly take up. Equally, an 'optimal' farm plan derived by linear programming methods – while it may indicate the ultimate opportunities for development – seldom provides a sufficiently realistic assessment of what will actually happen at the farm level when a project is in operation. A wrong assumption here may well distort the farmer's income picture and therefore his capacity to service his debt obligations. It may further overstate the incentive aspects of the project. Farm models, therefore, must be as fully representative as possible of average conditions in the project area.

The basic parameters from which the farm model is derived are as follows:

1 land use or cropping pattern
2 yields per unit area
3 output prices
4 physical inputs and costs of production per crop or enterprise
5 replacement costs
6 investment costs

Considerations to be taken into account in quantifying each of these, and a methodology which may be followed for integrating them in a dynamic model are briefly discussed below, accompanied by an example derived from a hypothetical project in Ruritania discussed later in this chapter. As will be noted, the model is developed to show the way in which the basic assumptions vary, as a result of the stimuli offered by the project, over time: from the pre-project situation until it can realistically be assumed that a point of stability will have been reached.

Land-use

The basic consideration in projecting land-use must, of course, be land capability, assessments of which can generally be based on the agronomic interpretation of soil surveys and associated maps. The degree of detail necessary in such surveys will vary according to both the intensity to which the land is to be used and the homogeneity of the soil. Where irrigated agriculture is to be developed, more detailed information about soils is necessary than in areas most suited, for instance, for low-intensity ranching. Soil surveys for a project area should show the location of the main classes of soil and demonstrate the suitability of each class for specific crops, indicating any special measures which may be required to reach the soil's best potential and overcome limiting factors.

Other physical factors determining the choice of cropping patterns include water availability and rainfall probability, temperature ranges and wind regimes. Frequently a project is designed to improve the availability of water through providing irrigation and in this case major changes between the pre-project and post-project cropping patterns may be predicted. The occurrence of frosts, or the danger of high winds – such as hurricanes in the West Indies – may limit the selection of possible crops.

Other non-physical constraints may be imposed on the cropping pattern. For agronomic reasons it may be necessary to restrict the frequency at which a particular crop (e.g. tobacco) may be grown on the same ground, and proven rotations may have to be followed to prevent the build-up of diseases and pests.

Marketing and price considerations may also limit the extent to which high-value crops are included in projected cropping patterns. Particularly on small farms, which generate relatively low cash surpluses, it may be necessary to limit the extent of saleable crops in the land-use plan and include a high proportion of crops for subsistence consumption, in order to reduce the risks to which a farmer exposes himself. In such cases, measures to guarantee and augment the farmers' own food supplies should have precedence over those to increase production for sale, although this should be adequate to enable each farmer to generate a cash income more than sufficient to service any debts incurred. Under the same physical conditions the projected land-use pattern could thus be expected to differ according to the size of the farm.

Other factors to be taken into account in proposing land-use patterns include the farmer's capability and prior experience with the proposed enterprise. This has greatest relevance with crops grown for subsequent processing where quality is a major factor. An example would be tobacco, where the price of the product is determined principally by its quality. The availability of manpower, either family labour resources or hired labour, is also a critical determinant in the choice of land-use patterns. Care must generally be taken to avoid a peaking of manpower requirements in a particular season.

A consideration which may be external to the farm enterprise relates to government priorities. It will be recalled that projects are sometimes viewed as instruments of public policy to orient a country's agricultural production. An interesting case in point is cotton in Burma. In pursuance of governmental policy to make the country less dependent on the imports of cotton manufactures the government, with West German bilateral assistance, established a modern textile manufacturing plant at the cost of some US$50 million. Established in 1972, the plant was not receiving enough cotton to run even one full shift by 1973. The spinning and weaving section was designed originally to use 132 000 bales of cotton annually, working two shifts. In order to meet the plant's raw cotton requirements, farmers were required by the village people's councils to grow up to 40 per cent of their cropped area to cotton. For this, the government provided machinery services for cultivation and agricultural credit. It was only in 1974 that the procurement price was raised to a level that farmers considered reasonable. By law, all long-staple cotton has to be sold to the government's Agricultural Corporation.

This case illustrates how the objectives of governments may run contrary to the private interests of farmers. Cotton, grown in the dry season in the existing irrigation tract, competed with other more profitable crops for limited water supplies and other farm inputs. For example, net returns per ha for medium-staple cotton of Kyats 885 compared with Kyats 1 418 for rainfed chillies, Kyats 3 729 for irrigated chillies, and Kyats 8 322 for onions. Farmers admitted that, while they gave the cotton crop two irrigations, the chillies and onions were irrigated eight times, and furthermore that inputs such as fertilisers intended for cotton, were diverted to the chillies and onions. These practices thus further tended to reinforce the persistence of low cotton yields.

Yields

Most agricultural projects have yield-increases as a basic objective. Assumptions about prevailing and projected yield-levels are critical to both the farm-level profitability (financial analysis) and to the economic benefits claimed for the project. Yet the methodology for projecting yields is far from developed, let alone understood by most project planners. Problems in yield measurement stem from the fact that yield is a function of the interaction of several different factors operating simultaneously. These include the level of technological knowledge; degree of control over the environment (e.g. water), institutional aspects (such as land tenure); the degree of farmer motivation and input–product price relationships.

Generally, the planner looks to the research stations as one source for guidance on levels of yields which might reasonably be expected. It is, however, difficult to bridge the gap between yields attained under conditions at research stations and those which would be expected under conditions prevailing on the farmers' fields.

This difficulty is accentuated because research programmes generally emphasise simple techniques (e.g. new varieties, fertiliser, time of planting, etc.) but rarely ever examine the complete farming systems to which they must be applied. This deficiency needs to be seriously examined by research workers in developing countries. The projection of yields for livestock production is considerably more complicated than for crops. The products of a dairy farming enterprise, for instance, include not only milk but also calves, some of which would be sold at varying ages, as breeding stock (e.g. in-calf heifers), fat stock (e.g. fat steers) or as young calves for veal.

An alternative source of information on which to base yield estimates is frequently a survey of existing farms, in which a separate analysis is made for progressive, average and poor farmers. A difficult part of the yield-projection exercise is forecasting the farmer's behaviour, i.e. his level and rate of response to technical assistance, credit or market stimuli. The farmer will adopt the new technology at a rate consistent with his appraisal of the benefits, and his capacity to secure and manage resources. It is in this context that the degree of technical support, particularly extension and other services provided by the project, plays a determining role in the progression of yield increases. This requires, therefore, that the system be appraised as a whole: for example, will inputs be available on time and at the right distribution points, to meet the farmer's requirements? The yield projection must take account of how major individual project actions, such as variety, water management, input package, extension services, credit and others, interact between themselves to affect yield.

Obviously, in trying to estimate project benefits, the initial or without-project yield assumption is as important as the projection of future yield. The planner's task is to collect and evaluate data on the existing situation and establish whether there is a built-in bias to inflate production and yield-data. This practice is not uncommon where special government-sponsored crop-production programmes are underway and jobs are endangered if reported results fall below expectation. Surveys may thus have to be undertaken of farms of representative sizes in the project area, based on representative crop years or crop seasons.

Output prices

Gross farm income may be projected by multiplying the area of each crop by projected yields for that crop and by the price received by the farmer for each unit of production. In the preparation of farm models conventional practice attributes actual or current farm-gate prices to all production, whether it is sold or not. Subsequently, in the preparation of the cash flow for the project, adjustments may be made for subsistence consumption by project beneficiaries.

Records of prices are maintained in most countries and analysis of these can provide a reasonable basis for estimating prices likely to be received by farmers for production in certain seasons. If price statistics are based on market surveys, these should be adjusted to the farm-gate level, allowing for transport costs and commissions. In estimating output prices, it is conventional to interpret price statistics with caution and conservatively.

Production costs

Farm operating costs may be derived from aggregating the direct or variable costs associated with producing each of the various crop or livestock products handled by the enterprise, and by adding appropriately calculated overhead costs. In the financial analysis, prevailing input prices (i.e. adjusted for subsidy or taxation) are those normally used. It is convenient also to account for all labour, whether or not this is supplied by the family, as though it was hired and to adjust the cash flow (discussed below) to take account of both family labour contributions (which will differ significantly between farm sizes) and subsistence consumption.

Replacement costs

The distinction is usually made in preparing the farm model, between normal operating costs and the costs of replacing machinery and equipment in the year in which these would be fully depreciated.

Investment costs

It is by stimulating investment at the farm level that most agricultural projects make their impact on production. Hence the choice of investments to be made in the context of the farm model is a critical element in project preparation. Decisions must be made as to what level of capital intensity the project should support, and what level of technology is appropriate for the typical farmer. The distinction must also be made between public investment (such as in main irrigation canals and diversion weirs, the costs of which may be recovered over a long period in the form of water dues) and strictly on-farm investment for which the farmer is wholly responsible.

Important decisions must also be taken on the phasing of investments. A frequent error in project planning is to anticipate an unrealistically rapid rate of investment. The amount of time which a small farmer can devote to development work in any one year is limited by his commitments to maintaining existing enterprises. The higher the rate of investment, the more the farmer is likely to have to depend on outside assistance and the lower is his proportionate contribution to development in the form of family labour.

When a package of investment items has been selected as the most appropriate for the particular type of farm, each item should be specified in considerable detail and costs should be estimated. Recent quotations or normally prevailing unit costs should furnish the basis for these estimates. All investment costs, including the farmer's own contribution to development, should be included in the estimates.

Cash-flow analysis

The interaction of income and cost streams, before and after debt service, and their relation to the farmer's debt-servicing obligations is demonstrated by means of a projected cash flow. The cash flow provides a convenient method of illustrating in a single table the following items.

The financing plan A summary of the respective contribution of the farmer himself, the lending agency and any other sources (e.g. shareholders, government grants) to financing the investment programme.

Projected income, investment costs, operating and replacement costs These are derived from the tables proposed above.
Debt servicing The obligations of the farmer under his loan agreement with the financing agency. Lending terms, where years of grace and of repayment should be specified and should be compatible with national banking practice and law as well as with the capacity of the farmer to service the debt out of increased income. Projections of debt service may be made either in the aggregated form of an annuity (which is convenient when the enterprise financed reaches a plateau of production rapidly) or as separate principal repayment and interest terms.
Net balance after debt service The level of income of the farmer after paying all costs including loan amortisation. If this is not significantly above pre-project levels even in the initial years of a project, there may be insufficient incentive for farmers to embark on the proposed programme, particularly if this exposes the farmer to greater risk, as it usually does.
Adjustments A true picture of the real cash position of the farmer may be obtained if the net balance after debt service is adjusted upwards by the estimated value of unpaid family labour and work off the farm, and downwards by the value of food consumed on the farm for subsistence purposes and hence not sold. Further adjustments may be made for taxes.

The cash flow provides the basic data on which to carry out a number of analyses. The most frequent analysis made involves the calculation of the internal rate of return which is based on the discounted incremental net balance before debt service. Possibly of more significance to the farmer is the assessment of the rate of return on his own investment after all changes, including debt service, have been covered. An additional figure of interest, especially for farmers with off-farm employment opportunities, is the estimated income per day of farm family labour.

A dairy farm and milk plant model

To illustrate the observations made above and to demonstrate conventional methodologies used in farm-model preparation, an example is given below of a typical farm model. The case used is of a 6-ha farm in Ruritania upon which a project sought to promote specialised dairy production (using high-yielding cows, better housing and mechanised fodder production) as part of a project which also included the expansion of an existing dairy processing plant and its related distribution system.

The dairy farm model represents a typical farm of 6 ha. Tables 10.1, 10.2, 10.3, 10.4, 10.5 and 10.6 show the progressive steps involved in deriving the cash flow and financial rate of return.

Table 10.1 shows first the proposed changes in cropping pattern on the farm. The most significant change introduced by the project would be the introduction of maize for silage. This, together with purchased concentrates would about triple the total feed units available. It should be noted that the value of production, without the project and with the project are expressed in thousands of currency units which are the units of account of the country.

Table 10.2 shows the projections for herd development, including

TABLE 10.1 Ruritania: Private sector, dairy farm (6 ha) – Cropping pattern, yields, production of feed units, and cash crops, value of production

Item	**Cropping pattern** (ha)		**Yield** (tonne/ha[1])		**Production** (tonne)		**Feed value** (feed units '000)		**Value of production** (units '000)	
	without-project	**with-project**	**without-project**	**with-project**	**without-project**	**with-project**	**without-project**	**with-project**	**without-project**	**with-project**
Grass (hay and grazing)	3·8	3·2	5	8	19·0	25·6	9·5	12·8	–	–
Maize for silage	–	1·6	–	40	–	64·0	–	13·0	–	–
Maize for grain[2]	1·2	0·8	3	5	3·6	4·0	–	–	9·0	10·0
Vegetables[3]	0·2	0·4	–	–	–	–	–	–	8·0	20·0
Wheat[4]	0·8	–	2	–	1·6	–	–	–	4·8	–
Total	6·0	6·0					9·5	25·8	21·8	30·0
Purchased concentrates							2·3	11·3		
Total feed units available[5]							11·8	37·1		

[1] As hay equivalent.
[2] At 2·5 units/kg.
[3] Gross value of production of vegetables estimated at 40 000 units/ha without-project and at 50 000 units/ha with-project.
[4] At 3·0 units/kg.
[5] The feed units available per cow (plus followers) are: without-project 2 350 and with-project 3 710.

TABLE 10.2 Ruritania: Private sector, dairy farm (6 ha) – Herd-development projections

Class of stock	Without-project	Year 1	Year 2	Year 3	Year 4	Year 5	Year 6	Years 7–15
Cows	5	5	7	8	8	9	10	10
Heifers first calving	1	4	3	2	3	3	2	2
Cows culled	1	2	2	2	2	2	2	2
Total end of year	5	7	8	8	9	10	10	10
Heifers 2–3 years	1	1	2	3	3	4	4	5
Sales	–	–	–	–	–	2	2	3
Total end of year	1	1	2	3	3	2	2	2
Heifers 1–2 years	1	2	3	3	4	4	5	5
Sales	–	–	–	–	–	–	–	–
Total end of year	1	2	3	3	4	4	5	5
Female calves	2	3	3	4	4	5	5	5
Sales	–	–	–	–	–	–	–	–
Total end of year	2	3	3	4	4	5	5	5
Male calves	2	3	3	3	4	4	4	4
Sales	2	3	3	3	4	4	4	4
Total end of year	–	–	–	–	–	–	–	–
Total AU[1]	7·3	9·4	11·8	13·9	15·1	16·4	17·5	17·5
Technical parameters[2]								
Culling rate (per cent)	20	40	30	25	25	20	20	20
Weaning rate (per cent)	80	80	80	90	90	90	90	90
No. of cows in milk	4	6	6	7	8	9	9	9
Purchases of heifers in calf	–	3	2	–	–	–	–	–

[1] Animal units: cows = 1; heifers 2–3 years = 0·8; heifers 1–2 years = 0·6; calves = 0·3.
[2] No mortality allowance included in view of high culling and insignificant effect on composition of small herds.

TABLE 10.3 Ruritania: Private sector, dairy farm (6 ha) – On-farm investment

Item	Year 1	Year 2	Total	Total	Foreign exchange	
	(units/'000)			(US$ '000)	(per cent)	(US$ '000)
Building and fixed equipment[1]	120·0	–	120·0	6·7	15	1·0
Mechanisation[2]	60·0	–	60·0	3·3	60	2·0
Livestock[3]	54·0	36·0	90·0	5·0	20	1·0
Total	234·0	36·0	270·0	15·0	27	4·0

[1] Average, allowing for adaptation of existing buildings.
[2] Equipment to be shared between 5–10 farmers depending on machine type; includes 35-hp tractors, forage harvesters, trailers, ploughs, maize seed drills, fertiliser spreaders and mowers.
[3] Purchase of in-calf heifers: 3 in the first year and 2 in the second year at 18 000 unit/head.

TABLE 10.4 Ruritania: Private sector, dairy farm (6 ha) – Operating costs (units/'000)

Item	Without-project	Year 1	Year 2	Year 3	Year 4	Year 5	Year 6	Year 7
Livestock								
Concentrates[1]								
Cows	1·5	12·6	14·4	14·4	16·2	18·0	18·0	18·0
Female calves	0·6	0·9	0·9	1·2	1·2	1·5	1·5	1·5
Male calves	2·4	3·6	3·6	3·6	4·8	4·8	4·8	4·8
Heifers	2·4	3·6	6·0	7·2	8·4	8·4	9·6	9·6
Milk or milk substitute: calves[2]	8·9	4·8	4·8	5·6	6·4	7·2	7·2	7·2
AI, veterinary services, animal health[3]	3·3	4·2	5·3	6·3	6·8	7·4	7·9	7·9
Electricity and water[4]	1·5	1·9	2·4	2·8	3·0	3·3	3·5	3·5
Maintenance and repair (machine and buildings)[5]	4·0	11·0	11·0	11·0	11·0	11·0	11·0	11·0
Insurance[6]	–	5·6	7·1	8·3	9·1	9·8	10·5	10·5
Sub-total	24·6	48·2	55·5	60·4	66·9	71·4	74·0	74·0

Crops								
Cultivation[7]	15·0	20·0	20·0	20·0	20·0	20·0	20·0	20·0
Seed[8]	7·6	8·2	4·4	4·4	8·2	4·4	4·4	8·2
Fertilisers: maize[9]	0·6	4·9	4·9	4·9	4·9	4·9	4·9	4·9
grass[10]	1·7	5·3	5·3	5·3	5·3	5·3	5·3	5·3
other[11]	0·5	–	–	–	–	–	–	–
Miscellaneous	2·0	4·0	4·0	4·0	4·0	4·0	4·0	4·0
Sub-total	27·4	42·4	38·6	38·6	42·4	38·6	38·6	42·4
Total costs	52·0	90·6	94·1	99·0	102·3	110·0	112·6	116·4

[1] At 3 unit/kg; cows 600 kg/cow (without-project 100 kg/cow); female calves 100 kg each; male calves 400 kg each; heifers average of 400 kg each.
[2] 500 litres of milk/calf without-project and 40 kg of milk substitute/calf at 20 unit/kg with-project.
[3] At 450 unit/AU.
[4] At 200 unit/AU.
[5] 10% of machinery capital costs plus 5% of building capital costs.
[6] At 600 unit/AU.
[7] At 5 000 unit/ha; 3 ha without-project and 4 ha with-project.
[8] Maize cord 150 unit/ha; maize silage 200 unit/ha; hay (every three years) 1 200 unit/ha; wheat 1 000 unit/ha; vegetables 10 000 unit/ha.
[9] Without-project 200 kg/ha of NPK at 3 unit/kg; with-project 700 kg/ha of NPK at 2·5 unit/kg plus 300 kg/ha of 27% at 1·4 unit/kg.
[10] Without-project 200 kg/ha of NPK at 2·3 unit/kg; with-project 600 kg/ha of NPK at 2·3 unit/kg plus 200 kg/ha of 27% N at 1·4 unit/kg.
[11] Without-project 200 kg/ha of NPK at 2·3 unit/kg.

TABLE 10.5 Ruritania: Private sector, dairy farm (6 ha) – Total value of production (units '000)

Item	Without-project	Year 1	Year 2	Year 3	Year 4	Year 5	Year 6	Years 7–15
Livestock								
Milk								
Milk yield/milking cow (litre)	2 000	2 400	2 700	2 800	2 900	3 000	3 100	3 200
Total milk production (litre)	8 000	14 400	16 200	19 600	23 200	27 000	27 000	28 800
Value of milk[1] ('000 unit)	35·7	64·2	72·3	87·4	103·5	120·4	124·4	128·4
Meat ('000 units)								
Cull cows[2]	9·9	19·8	19·8	19·8	19·8	19·8	19·8	19·8
Heifers 2–3 years[3]	–	–	–	–	–	30·0	30·0	45·0
Calves[4]	10·4	15·6	15·6	15·6	20·8	20·8	20·8	20·8
Sub-total	20·3	35·4	35·4	35·4	40·6	70·6	70·6	85·6
Value of total livestock Production	56·0	99·6	107·7	122·8	144·1	191·0	195·0	214·0
Crops	21·8	30·0	30·0	30·0	30·0	30·0	30·0	30·0
Total value of production	77·8	129·3	137·7	152·8	174·1	221·0	225·0	244·0

1 At 4·46 units/litre.
2 550 kg per animal at 18 units/kg.
3 At 15 000 units/each.
4 200 kg per animal at 26 units/kg.
5 Excluding value of forage used for on-farm animals consumption.

TABLE 10.6 Ruritania: Private sector, dairy farm (6 ha) – Cash-flow projections (units '000) **and financial rate of return**

Item	Without-project	Year 1	Year 2	Year 3	Year 4	Year 5	Year 6	Years 7–8	Year 9	Years 10–15
Cash inflow										
Value of production	77·8	129·3	137·7	152·8	174·1	221·0	225·0	244·0	224·0	224·0
Long-term loan[1]	–	163·8	25·2	–	–	–	–	–	–	–
Farmers' contribution[2]	–	70·2	10·8	–	–	–	–	–	–	–
Total cash inflow	77·8	363·3	173·7	152·8	174·1	221·0	225·0	244·0	244·0	244·0
Cash outflow										
Investment costs	–	234·0	36·0	–	–	–	–	–	–	–
Operating costs[3,4]	52·0	90·6	94·1	99·0	109·3	110·0	112·6	114·0	114·0	114·0
Replacement costs	–	–	–	–	–	–	–	–	60·0	–
Total cash outflow	52·0	324·6	130·1	99·0	109·3	110·0	112·6	114·0	174·0	114·0
Cash balance before debt service										
Debt service										
Loan outstanding	–	163·8	189·0	189·0	189·0					
Long-term loan amortisation	–	–	–	–	–	77·3	77·3	77·3	–	–
Interest on long-term loan (11%)	–	18·0	20·8	20·8	20·8					
Total debt service	–	18·0	20·8	20·8	20·8	77·3	77·3	77·3	–	–
Cash balance after debt service	25·8	20·7	22·8	33·0	26·0	33·7	35·1	64·3	70·0	130·0

Financial rate of return 22%

[1] 70% of investment cost.
[2] 30% of investment cost of which 20 000 units as labour contribution in the first year.
[3] Excluding family labour used for milking, feeding, etc.
[4] From years 7–15 an average figure has been applied.

the technical parameters, such as culling and weaning rates. The number of milking cows would, over a period of five years, increase from five to ten by buying five heifers during the first two years of the project.

Table 10.3 shows the planned on-farm investment involving holdings and fixed equipment, mechanisation equipment and purchases of in-calf heifers. This would total US$15 000 per farm. Investments would include the construction or adaptation of farm holdings to enable them to in-winter the additional stock under hygienic conditions.

Table 10.4 shows the derivation of operating costs. Total operating costs would increase from 52 000, without-project, to 116 400 currency units by the seventh year of operation. It should be noted that the technical assumptions for derivation of these costs are clearly shown. For example, the concentrate feeding per cow would increase from pre-project levels of 100 kg to 600 kg as the genetic potential of the cattle to respond to intensive feeding improves. It is useful to note in this regard that a fixed capital investment of currency units 15 000 generates an additional requirement for working capital of about currency units 63 000.

Table 10.5 shows the total value of production for milk, meat and crops expressed as thousands of currency units. Thus, average milk production would increase from 2 000 litres per cow per annum to 3 200 litres per cow per annum at full development in the seventh year. These increases would result from breed-improvement practices, involving artificial insemination, and improved feeding practices, both from better pastures and increased feeding of concentrates.

Lastly, Table 10.6 shows the cash-flow projections and the financial rate of return. The total cash inflow comprises the income from operations, expressed as value of production; the long-term loan from the project, and the farmers' contribution to project costs. The cash outflow shows the investment costs, and the annual operating costs, and the replacement costs of equipment in year nine. The debt service shows the loan amortisation and interest payments. Finally, the last column shows the cash-balance situation projected annually over the project. Thus, total fixed investments are estimated at 270 000 currency units, and net income is projected to increase from pre-project levels of 26 000 currency units to 130 000 currency units after completion of debt servicing and replacement cost by year ten. The financial rate of return works out at 22 per cent.

Financial analysis at other levels

The methodology described and illustrated above for the preparation of farm models may be applied with minor variations to provide a basis for financial analysis at other levels of a project. In the case of agro-industries, for example, a summary of technical coefficients indicating product mix and input–output relationship would replace the tables illustrating cropping patterns, yields and operating costs employed in the farm model. This is illustrated below, using the example of the milk-processing plant, referred to above, for which extensions were planned to accommodate the increased production of dairy products arising from the investment in dairy farms, based on the foregoing model.

TABLE 10.7 Ruritania: Milk plant – Summary of total investment costs and phasing

Item	Year 1	Year 2 (units '000)	Year 3	Total	Total (US$ '000)	Foreign exchange (per cent)	Foreign exchange (US$ '000)
Enlargement of the milk plant	–	1 300	17 300	18 600	1 016	73	745
Installation of a UHT/long life milk plant	25 000	–	–	25 000	1 365	69	944
Cheese-making plant	7 800	37 500	–	45 300	2 478	56	1 397
Setting up of two regional bulking centres	–	12 300	12 300	24 600	1 348	31	418
Transportation equipment	6 340	14 315	14 315	34 970	1 923	27	521
Cheese-making from sheep milk – village cheese plants	–	725	725	1 450	79	42	33
Storage/packing centre for cheese made from sheep milk	–	–	7 400	7 400	402	25	100
Total	39 140	66 140	52 040	157 320	8 611	48	4 158

TABLE 10.8 Ruritania: Milk plant – Income from sales

Item	Unit	Unit price	Quantity (units)			Value (units '000)		
			Without-project	Year 4	Years 12–20	Without-project	Year 4	Years 12–20
Pasteurised milk	litres '000	5 568	13 000	19 000	25 000	72 384	105 792	139 200
Yoghurts	litres '000	9 060	3 680	6 400	15 000	33 341	57 984	135 900
Cream	litres '000	28 000	1 380	1 400	1 800	38 640	39 200	50 400
Fresh cheese	tonnes	26 800	160	500	500	4 288	13 400	13 400
UHT milk	litres '000	7 000	–	7 300	7 300	–	51 100	51 100
Trapist cheese	tonnes	57 000	–	500	500	–	28 500	28 500
Fetta cheese	tonnes	40 000	–	375	375	–	15 000	15 000
Vlasic cheese	tonnes	45 000	–	200	200	–	9 000	9 000
Total						148 653	319 976	442 500

TABLE 10.9 Ruritania: Milk plant – Operating costs

Item	Unit	Unit price	**Quantity** (units)			**Value** (units '000)		
			Without-project	**Year 4**	**Years 12–20**	**Without-project**	**Year 4**	**Years 12–20**
Materials								
Cow's milk	litres	382	17 800	43 825	58 425	67 996	167 412	211 724
Sheep's milk	litres	800	–	700	700	–	5 600	5 600
Cream	litres	20 000	878	200	–	17 560	4 000	–
Packing materials						11 602	29 623	51 328
Water						534	1 315	1 889
Electricity						1 068	2 630	3 778
Fuel						1 068	2 630	3 778
Weight loss						1 120	2 850	3 755
Tyres						1 300	2 510	2 900
Service materials						220	300	340
Labour						3 560	8 765	12 593
Buying expenses						3 560	8 765	12 593
Distribution expenses						5 340	13 150	18 892
Repair, maintenance						1 643	4 045	5 811
Laboratory						1 080	2 658	3 820
Insurance						640	1 580	2 266
Travel allowances						1 268	3 120	4 483
General overhead						3 560	8 765	12 593
Other						1 629	4 850	8 106
Total						124 748	274 568	366 249

TABLE 10.10 Ruritania: Milk plant – Cash flow projection (million units) **and rate of return**

Item	Without-project	Year 1	Year 2	Year 3	Year 4	Year 5	Year 6	Year 7	Year 8	Year 9	Year 10	Year 11	Years 12–14	Years 15–20
Cash inflow														
Sales	149	163	216	283	320	335	349	363	377	391	405	415	443	443
Long-term loan[1]	–	31	53	42	–	–	–	–	–	–	–	–	–	–
Borrower's equity[2]	–	8	13	10	–	–	–	–	–	–	–	–	–	–
Total inflow	149	202	282	335	320	335	349	363	377	391	405	415	443	443
Cash outflow														
Investment costs	–	39	66	52	–	–	–	–	–	–	–	–	–	–
Operating costs	125	137	184	241	275	286	297	307	317	329	338	344	366	366
Replacement costs	2	–	–	–	–	–	–	6	14	14	–	–	37	6
Total cash outflow	127	176	250	293	275	286	297	313	331	343	338	344	402	372
Cash balance before debt service	22	26	32	42	45	49	52	50	46	48	67	71	40	71
Debt service														
Loan outstanding	–	31	86	135	146									
Long-term loan amortisation	–	–	–	– }										
Interest on long-term loan (8%)	–	(2)	(7)	(11) }	20	20	20	20	20	20	20	20	20	
Interest on working capital[3]	1	1	1	2	2	2	2	3	3	3	3	3	3	3
Total debt service	1	1	1	1	22	22	22	23	23	23	23	23	23	3

Cash balance after debt service and before taxes	21	25	31	40	23	27	30	27	23	25	44	48	17	68
Taxes	5	5	6	9	12	13	14	14	14	14	14	14	15	19
Cash balance after taxes	16	20	25	31	11	14	16	13	9	11	30	34	2	53
Financial rate of return 17·6%														
Rate of return on own equity 47·5%														

[1] 80% of investment costs.
[2] 20% of investment costs.
[3] 1 month turnover at 10% interest.

Table 10.7 summarises the capital investment costs involving plant expansion, the installation of new facilities, and transport and storage facilities at a cost of 157·3 million currency units equivalent to US$8·6 million dollars.

Table 10.8 shows that the income from sales of the various products – milk, yoghurt, cream and cheese – would increase from a level of 148·6 million currency units before the project to 442·5 million currency units at full development.

Table 10.9 shows the operating costs which would increase threefold and Table 10.10 shows the cash-flow projection and the rate of return. It is noted that the milk-plant enterprise would make a 20 per cent contribution to the project's investment costs, resulting in a 47·5 per cent rate of return on its own equity, in contrast to 17·6 per cent financial rate of return to the total additional funds directed to the project.

Commercial aspects

The term, commercial aspects, covers the activities related to the supply of goods and services required for the project's construction and operational aspects on the one hand, and to the commercialisation or marketing and distribution aspects of the project's output on the other hand. Such matters as input supplies and procurement are involved.

Illustrative examples of arrangements for raw materials and services required for the project are given below. Assume, for example, that the main investment in an asparagus-processing project would comprise the financing of the plant and equipment and warehouse facilities, with possible provision for the plant to supply a small part, say a quarter, of its raw-product requirements from its own asparagus-growing operations, with the remaining three-quarters to be supplied by independent outgrowers. The commercial aspects would include all the provisions to ensure that the raw materials and services required would be guaranteed to enable the plant to operate at its planned capacity.

With regard to the supply of raw asparagus, a system of contracts may be necessary, covering such items as varieties to be grown; cultivation practices to be adopted including spraying; time of planting; delivery times and quantities; quality specifications covering, for example, colour and texture. The contract would also cover the farm-gate price or the factory-delivery price, either through specifying price or through agreement on the formula for price determination. Other raw materials, such as sugar or salt, may be required, and the costs and sources of these inputs must be identified and assured.

The plant may have to purchase services, electricity, water and fuel. If the plant has to purchase electricity from an existing power network, the quantity required and the reliability of the source of power supply must be established, and costs determined. The same applies to water. Will water be available from existing water services, or will supplementary water supplies, possibly involving digging of its own wells be required?

Project works have to be constructed. What arrangements would be made for selecting required equipment, for transport and assembly at the plant site? What would be the role of contractors, consultants, equipment dealers? What arrangements exist for supervision of work of contractors and consultants, and are their terms of reference clearly spelled out? Will international competitive bidding be fully adhered to or will there be preferential clauses favouring local contractors?

In the more usual agricultural project, e.g. irrigation development, some of the problems that arise with regard to inputs and services are as follows. Are fertilisers, insecticides, hand-tools available in the project area in the quantities and at the times required for farming operations? What distribution services exist? Are there well-established dealer networks, or are fertilisers and seeds, for example, obtained by the farmer from local merchants? Or is there a government agency which distributes inputs, and do they have adequate storage facilities and transport for delivery to various villages? If not, the project should ensure the provision of such facilities, for example, by financing private contractors to undertake cultivation work, or should it reinforce any existing services, where necessary? Under what terms and conditions are such inputs available to farmers? Are interest rates exorbitant?

Where the project provides a mechanisation component, it is important to ascertain the availability of spare parts and repair services in or around the project area. A condition of procurement of equipment might well require the dealer to establish a service and repairs unit in or near the project area. Alternatively, the project may have to establish its own repair-services unit and make the necessary provision for training of personnel in the operation and maintenance of the equipment.

Markets and distribution

The project's output must be marketed. The project planner's task is to ensure that any deficiencies along the existing marketing chain of the product do not constitute a constraint. The structure of marketing must be determined: is it a monopoly or an oligopoly, or is it a competitive market? Is there a strong marketing board that actively intervenes, or does it play the role of residual buyer? The existence and performance of producers and marketing associations is important. A further major problem arises where there is an inefficient publicly supported marketing operation, operating side by side with an efficient private enterprise.

Where an already established channel and a strong market demand exist, then no project action is required. However, the situation not infrequently arises where the product's market would have to be developed. In a seeds project for example, it may be necessary to build a transport and distribution network. The project's organisation may well include a sales and promotion division or unit, amongst whose tasks may be the development of forward seed contracts for seed sales.

Where a country decides to enter production for export for a commodity in which it has little or no marketing experience, it may be necessary to develop linkages with internationally recognised firms which have their established distribution outlets. Examples are bananas and pineapples, where brand is important to the consumer. Vegetable production in the Middle East countries aimed at the European market is another example. In such difficult market situations, the project's organisation must ensure reliability of supply, and superior quality. Grades and standards take on enormous significance, and failure to meet acceptable standards can threaten the financial viability of the enterprise. In summary, the commercial requirements of the markets for the product must be reflected at the farm level.

References

1 J. Price Gittinger, *Economic Analysis of Agricultural Projects*, The Economic Development Institute, International Bank for Reconstruction and Development, Johns Hopkins University Press, Baltimore, 1972.

11 Economic and social aspects

Economic aspects

Assessment of project benefits

The justification of a project depends on the extent to which it can be shown to satisfy the broad development objectives of the sponsoring government and source of finance. It has been customary in the evaluation of projects to place greatest emphasis upon assessing a project's effect on the national economy, and tools of economic evaluation are progressively being refined to give a quantitative measure not merely of aggregrate benefits but also of the distribution of these benefits.

The need to meet other than purely economic objectives in project design is well recognised, and economists have been tempted to modify their analytical tools to arrive at 'magic' figures which indicate the extent to which a given project satisfies multiple objectives. Since the objective of assessing a project's benefit, however, is to provide decision-makers with a national basis for deciding whether or not to commit financing for its implementation, there are considerable advantages in treating a project's benefits in a disaggregated manner, distinguishing between those benefits that are essentially economic and those which relate to a project's impact on society. Among the latter it is probably useful to consider questions relating to the number of direct and indirect project beneficiaries according to their pre-project income levels. It may be useful also to review the impact of a project in relation to the following.

1 Employment creation.
2 Nutritional effects.
3 Rural – urban migration patterns.
4 Standard of living.
5 Environmental stability.
6 Shares of project-derived income according to the poorest groups in the project area.

Purposes of the economic analysis

The purpose of the economic analysis is to derive quantitative measures of a project's impact at the level of the national economy which can be taken into account, along with other considerations, in decisions about establishing priorities and approving finance.

This simply stated objective confronts very difficult application in practice. For example, a production project located in the more favourable areas of a country where soils, transport facilities and markets exist, probably would show a higher economic rate of return than one located in the more depressed areas of the country. In fact, these areas are depressed precisely because of an adverse environment, and a paucity of natural and human resources. If a major objective is the redressal of inter-regional or intra-regional disparities in income or employment or growth potential, then the planner probably has to settle for a lower economic rate of return in projects designed to redress this situation. The difference between the rate of return and that on other possible projects (where comparison can be made) indicates the cost of a decision to accommodate non-economic objectives.

The interpretation that should be given to a project's economic rate of return only makes sense, therefore, in a social context. There is hardly a project which does not have a social dimension to it, because options usually exist as to the location of project, the potential beneficiaries, and the income-distribution aspects of the investment. These decisions are tough, especially where a country may face serious food deficits and the urgency of the situation may require the fastest possible approaches to increased production.

Another dimension of the problem is the limitation of this analytical tool for all types of agricultural projects. Whereas, for example, projects in other economic sectors, e.g. industry, transport, tourism, might readily lend themselves to the measurement of output, that is, the benefits stream, this does not readily hold for agriculture. This applies in particular to some very basic and well-recognised aspects of the agricultural development process, such as research, agricultural extension, agricultural education and training.

It is well known that investment in agricultural research can have a quite high pay-off, but the measurement of economic benefits derived is difficult and speculative. The same applies to extension and agricultural-education projects. We require extension to make the farmer receptive to new technology but other factors besides sheer knowledge of the improved technology (e.g. product prices), may be much more significant in the rate of adoption than the extension input *per se*. Financing institutions, faced with the problem of measuring benefits, do finance agricultural education and research projects based on the overall justification for the project.

The same problem applies to the measurement of economic benefits of nutrition projects. Obviously a nutrition project should be so designed that the project's output is directed towards specific target groups of beneficiaries.

Joy and Payne[1] argue that objectives in nutrition are unlikely to be adequately reflected and are likely to be denied by cost-benefit procedures which use as their criteria the increment of aggregate net production valued at market prices, deriving from a project. Nutrition planners should reject this because the market does not reflect the social values which are the basis of national policy. Acceptable objectives might, however, be broadly reflected in procedures which show concern with the pattern of consumption between people. This could be expressed by noting the impact on the numbers in low-income groups, perhaps by assigning extra weight to increments of income to people in these groups. The acceptance of measures to improve nutrition will always depend on the values imposed by politicians and decision-makers.

Emerging analytical tools

Economists, recognising the limitations of the tools of economic analysis, are striving to refine the existing ones. The UNIDO guidelines[2] are one such instance of a sophisticated attempt to examine the basic issues of economic evaluation of projects. Hansen[3] has analysed and summarised the guidelines into a simple document that lists and discusses the steps in project evaluation in the UNIDO guidelines.

The Little and Mirrlees[4] study on social cost-benefit analysis is a well-known work in this field and a 1975 World Bank publication[5] represents the latest contribution to these new approaches for project evaluation.

Frequently in project formulation the economic analysis of agricultural projects is a tool employed *post facto* to justify a project already selected and designed. It is seldom now used as a tool in the earliest stages of project formulation to evaluate possible alternatives, and the impact that such alternatives may have on the development objectives of the country. In other words, present practice aims at establishing that the project selected has met the test of a minimum acceptable rate of return to the economy. It does not say to what extent the project contributes to the non-economic objectives of the government.

We attempt to set out briefly the type of thinking and broad approaches being developed to cope with the problem, before moving on to the conventional methods of analysis. It is the lack of confidence in the latter that has induced the search for new approaches.

The economic rate of return assesses the costs and benefits of a project and reduces it to a common denominator. Since it is a tool to measure the benefits to the economy deriving from the investment, then the economic rate of return for the selected project should be at least equal to and preferably exceed the opportunity cost of capital for the particular country.

Current practice aims at allocating various resources consumed by the project so as to maximise the contribution to the country's gross national product. The project generates an output stream whose value is measured employing 'efficiency prices'. These are tantamount to economic prices which assume implicitly that the weighting of the value of the distribution of the project's output amongst different beneficiaries is equal to unity.

This is possibly over-simplified because governments have different development objectives. These may include for example, an accelerated economic growth rate; increasing government revenues; building up foreign-exchange reserves, and other fiscal considerations; raising employment levels; increasing private consumption levels, or alternatively, raising the level of savings as a percentage of the GNP; and the distribution of income to correct for undue concentration of income among certain groups.

It is clear that these different objectives may conflict. Thus, a high rate of investment and savings implies deferring consumption for a later period instead of raising immediately current consumption levels. A project whose beneficiaries are primarily subsistence farmers would have a lower reinvestment quotient than one aimed at large-scale commercial producers. The types of consumption generated could be significantly different, with the former resulting in increased food intake while the latter results in consumption of luxury goods.

Faced with these conflicts, the situation requires decisions on trade-offs between the country's development objectives. Thus, the shadow-pricing of inputs must reflect the constraints posed by the availability of key resources – primarily capital, foreign exchange and labour (especially skilled labour) – in the developing countries. Apart from these more recognisable constraints, more intractable constraints may derive from other sources, e.g. the government may find it politically impossible or difficult to remove a long-standing subsidy, which leads to a distortion in the resources-allocation function.

On the output side, the task of the analyst is to measure the value of the output – a value that depends, however, on the distribution effects of the project – that is the share of income going to different groups of the society. The aim, therefore, of emerging analytical tools is that calculations of the economic rate of return must be adjusted to reflect the project's distributional impact. This is done by the application of a system that introduces criteria for the derivation of distribution weights.

These new measuring tools are still very much at the conceptual stage and still have to undergo considerable testing under the varying conditions in developing countries. Their limitations must be fully tested, before they become available for general use by project planners. While the use of these new methods would be much more time-consuming and require considerably more data for the analysis, yet they deserve to be fully supported in the search for analytical tools to ensure better project selection. The formulation of projects designed to meet the differing development objectives of governments must be furthered in every possible way.

Conventional economic analysis

Gittinger's[6] treatment of the subject warrants close attention. The preceding discussions on financial analysis in Chapter 10 covered the farm model, the farm budget, and cash-flow tables leading to the calculation of a financial rate of return.

The data from the foregoing frequently form the basis, after adjustment, for deriving the economic rate of return. Adjustment takes two forms, namely, removal of some cost items and inclusion of new ones, and secondly the attaching of weights or new values to the factors of production and the project's output through the mechanism of 'shadow pricing'. These are, in turn, discussed below.

The deciding factor, as to whether an item is included in the economic analysis or not, is whether it makes a direct claim on the resources of the country or whether it merely represents a transfer of ownership of the resource from one group or entity or individual to another. For example, a fruit cannery's payment of interest to a credit institution, for overdraft facilities to finance inventories, represents a transfer, and such interest payments are excluded from the economic analysis. Payments for land purchases are also generally excluded.

Taxes fall in the same category of transfer payments, be they customs duties, income taxes, sales or purchase taxes or even religious taxes. While impinging on the financial situation of the enterprise or the farmer, they do not represent new claims on the country's resources. The same applies to dividend payments to shareholders. While these are considered as costs in the financial analysis, they are not deducted from the income stream in the economic analysis.

In contrast, subsidies, as briefly mentioned earlier, are included in the economic analysis, since they represent a cost to the country as a whole and are part of the country's investment in resources allocated to the project. The subsidies can take several forms – interest rates below the market interest rate; purchased farm inputs; or mechanisation services provided at concessionary prices.

The economic analysis must include all the investment which is directly attributable to the project. Many of these costs may not appear in the financial analysis. Thus, all physical infrastructure, roads, ports, irrigation works; all incremental overhead-management costs, including staffing, housing, transport facilities; all technical assistance, local and expatriate; and all social services, e.g. clinics and schools, are a cost to the project. Additional support services, extension, electrification and marketing facilities, which would not have been undertaken in the absence of the project, should be costed.

It is not uncommon to find in practice that many of these items are forgotten. It is important to differentiate between what is a project cost and the economic costs of the project. The latter costs are more truly representative of the total investment resources of the country to which the project lays claim. To illustrate this, it is quite frequent to find that as a condition of loan negotiation, the financing institution or donor country may require the government to increase the extension services in the project area to a stated minimum ratio of extension worker to farm families. A local credit institution may be required to establish branch units in the area. These incremental services represent a real cost to the economy and should be included as such in the economic analysis.

Shadow pricing

Usually shadow prices are considered for, or applied to, three main items: labour, foreign exchange, and to the project's output where such goods are internationally traded. They are applied so as to correct for distortions in the applications of resources occasioned by market imperfections. The shadow price represents an attempt to establish the scarcity value or the equilibrium value that would have prevailed in the absence of such distortions.

With regard to labour, the shadow wage rate should be treated separately for unskilled labour and for skilled labour. For unskilled labour, the prevailing agricultural wage rate in the project area forms the basis for the financial analysis. In the more populous Asian countries it was sometimes assumed that the marginal value product of labour was zero or very close to it. This assumption naturally tends to put the project in a more favourable position as measured by the economic rate of return. Yet, in actual practice, the seasonal nature of agricultural production, largely dictated by monsoonal behaviour, means that there is a clumping of operations, and certain farm activities (e.g. transplanting paddy), must be undertaken within certain defined time periods, when the otherwise surplus labour pool approaches zero and shortages may in fact appear. During such periods, the marginal value of labour certainly lies above zero and moves up in the direction of the market wage rate.

With regard to skilled labour – managers, technicians, mechanics, etc. – the prevailing wage scale may well underestimate the economic cost to the society of such skills within the project. This is especially the case when government-administered projects are undertaken side by side with a dynamic private sector. The higher wage scales and perquisites in the latter may well be a better guide to the economic cost of such skills than the civil service emoluments.

Foreign exchange

The true value of a country's currency is the price that it would fetch in the open market if it were fully traded. In practice, fixed exchange rates exist where one country pegs its currency to another, such as the US dollar. Floating exchange rates are, however, becoming quite popular. Most countries have an official exchange rate and, in fact, may have multiple exchange rates, designed to accomplish various objectives. Thus, tourism may be encouraged by having a premium rate for tourists entering the country. The exchange rate could also be manipulated to favour certain segments of the economy, e.g. export industries may get favoured rates over producers for the domestic market, because the former are foreign exchange earners. Controls may exist on imports and movements of capital as governments introduce measures to protect their foreign exchange reserves.

Because of all these imperfections, the official exchange rate may well represent an over-valuation or under-valuation of the country's currency. For example, a black market rate for foreign exchange that is considerably above the official exchange rate suggests that the currency is over-valued.

The question thus arises of what is the appropriate shadow price to use in assessing the domestic currency equivalent of the imported investment items for the project. Gittinger has suggested that the project analyst should use the shadow price derived from the country's central planning unit, as every project would have to use the same figure, but there could be reluctance since this might imply official acceptance of over-valuation.

Market prices

While problems exist in trying to relate prices of domestically produced goods to those of internationally traded goods because of grades and quality factors, the margin of error is apt to be less than relying solely on domestic prices as an indicator of true economic prices, because of government's pricing and procurement policies or other market imperfections. For example, in Malaysia, the domestic rice price exceeded the international price in response to a policy of self-sufficiency and stimulation of domestic production. In Thailand, a reverse situation obtains with the domestic price lying below the international price, in pursuance of a policy of maintaining low consumer prices, and imposing export levies.

Additionally, governments may pursue procurement policies that distort the market prices. In Burma, for example, the government's price for paddy averaged Kyats 462 per tonne in 1975, while the free market price averaged Kyats 974 per tonne. The average financial farm-gate price for paddy worked out at Kyats 803 per tonne, in contrast to the economic price calculated at Kyats 1 015 per tonne, including the use of a shadow rate of exchange.

Incrementality

The economic analysis is a means of establishing the incremental benefits arising from the project's investments. Therefore, the analysis is concerned not with a pre- and post-project approach but with the 'without-project' and 'with-project' situation. That is, we cannot assume a static situation would prevail throughout the project area in the absence of the project, unless all the evidence points in that direction.

The production trend line may show three possibilities: a declining trend, a static one, and a slowly rising one. A declining one could exist, for example, because of continuing erosion or adverse pests and diseases, or because of market forces, the development of salinity or other causes. In such a case, the 'with-project' benefits are not only the incremental gains over the situation at the start of the project, but would also include the 'losses' that would have occurred in the absence of the project.

Sensitivity tests

In order to identify the main potential weaknesses in a project or to give decision-makers an idea of the risks of not attaining forecast rate of return, it is usual to test the sensitivity of the calculated economic rate of return to changes in cost and income assumptions. Sensitivity tests may be made, for instance, on the effect of a 10 per cent increase in capital costs and a 10 per cent decrease in product prices on project returns. Sensitivity tests also provide a useful means of assessing the economic cost of delays in construction and the initiation of the higher levels of production to be achieved as a result of embarking on the project.

Sunk costs

Sunk costs can be defined as costs already incurred in the project area where the new project investments will be directed. Since economic analysis concerns itself with the incremental benefits deriving from incremental investments under the project, then the existence of sunk costs can sometimes considerably boost the returns, thereby suggesting that the earlier works undertaken should be carried to completion. Irrigation completion projects are a case in point.

A broader issue arises, however, when several on-going large programmes or projects have to be completed to realise the full potential of investments already undertaken. This is the case for Command Area

Development Projects in India. With project works at all varying stages of development within the same state and among different states, a dependence on the economic rate of return to assess priorities between the various projects can lead to erroneous conclusions. This arises because the rate of return does not say which of the programmes and projects makes the most sense within the context of the country's overall irrigation development objectives. This requires, therefore, a broader overall look at the individual programmes and the totality of the programmes in relation to the total resources available for the sub-sector; the incidence of sunk costs would be one of the elements considered.

Benefits and justification in an irrigation project

This example, based on an irrigation project in a state in India, demonstrates how the various benefits of the project were assessed and how the economic rate of return was determined.

The objective of the project was to change from a drought-prone rainfed area to a highly productive irrigated area. The major project components consisted of irrigation and drainage works, roads, buildings, pilot water management, equipment and vehicles, and engineering and supervision. The project represented support for an on-going scheme in which the dam and spillway, and some of the primary distribution system had been completed.

The project's incremental production would consist of about 217 000 tonnes of food grains, mainly sorghum and wheat, and about 180 000 tonnes of cash crops, mainly sugar cane. The value added to the local economy would be Rupees 446 million.

The project's employment effects were assessed both in terms of its direct and indirect effects. The increase in family employment would represent about 12 600 full-time jobs.

The generated farm employment for non-family labour would be about 29 500 full-time jobs, the main beneficiaries of which would be landless labourers.

The indirect employment generated by this increase in value added in agriculture would be felt in local non-farm sectors such as marketing, transport and processing. The generated non-farm employment was estimated at 13 900 man-years on the basis of earlier studies in an adjoining state, which showed that an increase in value added in agriculture of Rupees 1·0 million would generate 60 full-time jobs in the non-farm sector.

The estimated income effects of the project's investment of a 5-ha farm would be an increase of 3·5 to 4·2 times, depending on the soil types, as the irrigation constraint is removed. The average income of 88 000 landless labourers would double as their average annual working days would increase from 130 to 230 days.

The project's impact on rural poverty would be substantial. Currently, about 56 per cent of farm families live below the poverty line of Rupees 600 per annum. The project would reduce this proportion to about 10 per cent.

The project's economic rate of return, including sunk costs represented by investments previously undertaken, was estimated at 13 per cent, assuming an economic life for the project of 50 years. However, if all sunk costs were excluded from the analysis, the economic rate of return would be about 33 per cent.

The sensitivity analysis showed the following for changes in assumption for key variables. A 25 per cent increase in the construction costs would drop the rate of return to about 11·0 per cent. A decrease in expected benefits of 20 per cent would also decrease the rate of return to 11·0 per cent, while a delay in full benefits of 5 years would also give a rate of return of 11·0 per cent.

Basic assumptions in deriving the rate of return (ERR)

The economic analysis is based on the use of economic costs for the project's investments and the use of economic prices in the measurement of project's benefits. This requires conversion of financial costs and prices to economic costs and prices.

The main assumptions were: The World Bank's projected 1990 world market economic prices for major traded agricultural outputs and inputs were employed in the analysis and expressed in 1979 currency values with appropriate adjustments for freight, handling and processing. For the non-traded food gains, the projected economic prices were derived by applying a food-grain conversion factor to the financial price.

With regard to the official exchange rate (OER), conversions of financial prices were made at the rate of US$1.00 to Rupees 8·50. The Indian rupee is now fixed against a 'basket of currencies' and since these currencies are floating, the US$/Rupee exchange rate is subject to change.

The standard conversion factor (SCF)

The benefits of the project must be evaluated at world market prices, that is, they measure what India would have to pay for imports or receive for exports. Most of the project's costs, however, involve items that are not traded in the world market. In order to make costs more directly comparable to benefits, it is necessary to adjust costs by applying a standard conversion factor to the price of non-traded and consumption goods. The Indian Planning Commission recommends for the evaluation of industrial projects the use of conversion factors for foreign exchange and for untraded goods that implies a SCF of about 0·75. For the economic analysis of this project, a SCF of 0·8 was used.

Prices of agricultural outputs

The economic prices for traded goods such as rice, wheat, cotton, groundnuts and sugar can be based on their border prices at the official exchange rate. The domestic cost component – including local transport and marketing charges – is adjusted by the SCF of 0·80. In contrast, for non-traded food grains such as sorghum, maize and pulses, the economic prices were derived by applying a food-grain conversion factor (FCF) to the financial price.

The following formula was used to estimate the FCF:

$$\text{FCF} = \frac{(\text{Q rice} \times \text{PE rice}) + (\text{Q wheat} \times \text{PE wheat})}{(\text{Q rice} \times \text{PF rice}) + (\text{Q wheat} \times \text{PF wheat})}$$

where, Q = the production of rice and wheat in the state; PE = economic price; and PF = financial price.

The projected financial prices for various crops were based on the estimated historical price trends in the state and in the country. The summary of prices for economic and financial analysis is shown in Table 11.1 for the years 1979 and 1990.

Prices of agricultural inputs

A conversion factor of 0·95 was used to derive the economic price of fertilisers, pesticides and insecticides from the projected financial price.

TABLE 11.1 India: Irrigation project – Summary of prices for economic and financial analysis (Rupees per tonne)

Crops	Economic prices 1979	Economic prices 1990	Financial prices 1979	Financial prices 1990
Sorghum – *kharif:* local	1 320	1 420	1 290	1 390
HYV	1 320	1 420	1 290	1 390
Paddy – transplanted	1 220	1 440	950	1 380
– drilled	1 370	1 610	1 070	1 540
Pearl millet – *kharif*	1 150	1 410	1 120	1 380
Maize – *kharif* and *rabi*	1 150	1 410	1 120	1 380
Groundnut – *kharif* and *rabi*	2 600	2 670	2 270	2 750
Cotton – local	4 070	4 370	4 000	4 300
– hybrid	5 060	5 440	5 000	5 400
Chillies	5 220	7 700	6 000	7 700
Turmeric	4 000	4 400	4 000	4 400
Onions/vegetables – *kharif* and *rabi*	600	600	660	660
Pulses	1 910	1 800	1 870	1 760
Pigeon pea *(tur)*	1 830	1 800	1 790	1 760
Wheat	1 270	1 530	1 500	1 650
Sorghum – *rabi:* local	1 440	1 540	1 410	1 510
HYV	1 320	1 420	1 290	1 390
cotton	1 320	1 420	1 290	1 390
Gram/peas	1 510	2 140	1 480	2 090
Potatoes	850	860	930	940
Safflower/linseed – rainfed and irrigated	2 330	2 700	2 180	2 530
Lucerne	200	200	220	220
Fodder – *kharif*, rainfed	100	100	110	110
– HW irrigated	120	120	130	130
Sugar cane	130	160	120	160

Thus, the financial price for nitrogenous fertilisers of Rupees 4.64 per kg was converted to an economic price of Rupees 4.41 per kg.

A shadow wage rate of 65 per cent of the market wage was assumed. In Indian irrigation projects, studies showed that the economic shadow wage rate for farm labour lay in the range of one-half to two-thirds of the average financial price of Rupees 4.0 per man-day. This was converted to an economic price of Rupees 2.6.

Construction conversion factor (CCF)

The main components of the project's construction costs include a mixture of equipment, skilled and unskilled labour. The traded component includes capital-intensive works which require imported materials. Since this is traded, a conversion factor of 1·00 was used for the estimated 15 per cent of the construction costs falling in this category.

The non-traded component includes works involving the use of skilled labour and locally manufactured materials. A standard conversion factor of 0·80 was used for this component, involving about 30 per cent of the construction costs.

For the unskilled labour component, about 55 per cent of the construction cost, a conversion factor of 0·65 was employed. The overall conversion factor based on the weighted average of the above three components worked out at 0·75.

The cost and benefit streams for the economic analysis – in 1979 economic prices – are shown in Table 11.2.

The analysis covers a 50-year period from 1965 to the year 2015. The analysis includes all the investment capital costs incurred since the project's inception in 1965 and the new capital costs involved in the continuation and expansion of the irrigated network and associated development. The addition of operation and maintenance costs gives total project costs.

Project benefits represent the economic value of increased agricultural production. The difference between this item and the total project cost for the same year gives a negative or positive balance shown under net project benefits. On the above assumptions, the project's economic rate of return works out at 13 per cent. Since this lies above the opportunity cost of capital in India, there is adequate economic justification for undertaking the project.

Limitations of the rate of return as an analytical tool: land-use in Nepal

Too frequently the economic rate of return of a project is accepted as the 'last word' in the analysis of a project's economic benefits. While in many cases it provides a perfectly adequate assessment of the likely impact of investments on the economy, there are other forms of economic analysis which may be quite as relevant in determining the effects of a project. For instance, two possible projects for the same area may generate similar rates of return but have completely different financing requirements and different levels of contribution to a country's gross domestic product. The applicability of other measures is illustrated by the following example,

TABLE 11.2 India: Irrigation project – Cost and benefit streams for economic analysis (1979 economic prices [million Rupees])

Year	Capital cost	Operation and maintenance cost	Total project cost	Project benefits	Net project benefits
1965/66	1·11		1·11		1·11
1966/67	1·46		1·46		1·46
1967/68	3·15		3·15		3·15
1968/69	5·77		5·77		5·77
1969/70	14·24		14·24		14·24
1970/71	25·01		25·01		25·01
1971/72	34·14		34·14		34·14
1972/73	52·85		52·85		52·85
1973/74	61·32		61·32		61·32
1974/75	63·90		63·90		63·90
1975/76	124·70		124·70		124·70
1976/77	117·05		117·05	1·76	115·29
1977/78	110·20	0·21	110·41	8·37	102·04
1978/79	125·63	0·68	126·31	19·37	106·94
1979/80	123·19	1·24	124·43	34·34	90·09
1980/81	100·31	1·80	102·11	55·47	46·64
1981/82	15·42	2·59	18·01	81·00	62·99
1982/83	45·44	3·49	48·93	111·82	62·86
1983/84	133·31	4·46	137·77	150·56	12·79
1984/85	129·62	5·44	135·06	195·90	60·84
1985/86	142·72	6·41	149·13	239·05	89·92
1986/87	143·48	7·39	150·87	279·99	129·12
1987/88	144·28	8·36	152·64	338·11	185·47
1988/89	132·61	9·34	141·95	377·29	235·34
1989/90	24·19	9·45	33·64	404·58	370·94
1990/91	23·66	9·45	33·11	422·19	388·08
1991/92	25·09	9·45	34·54	431·44	396·90
1992/93	25·53	9·45	34·98	438·04	403·06
1993/94	25·96	9·45	35·41	439·80	404·39
...	...	...	...	...	...
2014/15	26·40	9·45	35·85	440·24	404·39

which compares the benefits, at various levels, of different options for land development in the Nepalese Terai.

The area under consideration was covered by natural hardwood forest, and three possible courses of development could be adopted.

1 Intensifying management of the natural forest.
2 Establishing plantations of fast-growing forest species.
3 Establishing small-farmer agricultural settlements.

Relative economic rates of return

Calculations on the economic rates of return which would be generated by each potential type of development were made using assumptions shown in Table 11.3. No cost is attributed to unskilled family labour utilised on lessees' and settlers' farms but other labour costs are included at their financial value. Projected prices for agricultural goods are below current prices in the Bardia/Banke region. Standing forest timber is valued at Rupees 107 per m³ (net usable volume) since estimates show that given current lumber prices in the Indian market, this represents the approximate return to the country after making allowance for all costs associated with felling, extraction, transportation, sawmilling and distributing of timber produced in the area.

The estimated economic rates of return for each potential course of action are summarised in Table 11.3.

TABLE 11.3 Nepal: Terai – Economic rates of return for possible development alternatives

Action	Assumptions	Rate of return (per cent)
Intensified forest management	Immediate increase in extraction rates by 300 per cent. Subsequent doubling of annual cut after year 30.	11
Eucalyptus	1 10-year rotation with 4 years' initial cultivation by lessees	
	(a) Final yield 55 t/ha	31
	(b) Final yield 35 t/ha	29
	2 10-year rotation with no income from initial cultivation	
	(a) Final yield 55 t/ha	16
	(b) Final yield 35 t/ha	12
Land settlement	1 Settlers' net income Rs 1 500/ha	38
	2 Settlers' net income Rs 1 000/ha	25
	3 Settlers' net income Rs 500/ha	7

In view of the extreme range of rates of return associated with plantation establishment, depending on the extent of cultivation by lessees in the early years, a straightforward ranking of possibilities on the basis of internal rates of return fails to emerge. It is possible, however, to come to the following conclusions, assuming that return on investment would be a determining factor.

1 Intensified forest management (even under highly optimistic assumptions) cannot compete with the other two alternatives.
2 Plantation establishment is only likely to be a more paying course of action than land settlement if it is associated with a lessee cultivation scheme during the clearing and tree-planting phase.

If the development of a satisfactory system of lessee cultivation is not practicable, the maximum potential rate of return for plantation establishment would be about 16 per cent which would not compete with a return of 25 per cent projected for agricultural settlement using modest yield assumptions.

Investment and returns per unit area

In a country such as Nepal in which productive land is extremely limited in area, considerable importance must be attached to the intensity of

TABLE 11.4 Nepal: Terai – Cost/benefit ratios for possible development alternatives

Action	Discounted investment (Rupees per ha)	Discounted returns (Rupees per ha)	Discounted net returns (Rupees per ha)	Cost benefit ratio
Intensified forest management	388	415	27	1:1·07
Plantations (no lessees, 5·5 t/ha per year)	2 433	2 926	493	1:1·2
Agriculture (net sales Rupees 1 000 per ha per year)	5 460	8 335	2 875	1:1·5

development in a given area. A comparison is made in Table 11.4 between the capital costs and returns per hectare associated with each potential form of development. Both costs and revenues are discounted at 10 per cent per annum over 30 years.

From Table 11.4 the ratios below are derived.

Investment in agriculture to investment in plantations	2·25:1
Investment in agriculture to investment in forest management	14:1
Returns from agriculture to returns from plantations	2·84:1
Returns from agriculture to returns from forest management	30:1

The principal implication is that, whether capital is limiting or not, agricultural development provides greater gross and net returns than other development possibilities per unit area of land. If the assumptions employed above are accepted as representing realistic alternative methods of land-use in Bardia, it can be shown that net aggregate returns to the economy resulting from agricultural development are some six times greater than those from plantations, and over 100 times greater than those stemming from the intensification of forest-management activities in the area.

If the feasibility of developing irrigated agriculture was to be demonstrated, the returns to agricultural development per unit area would almost certainly be even greater, relative to those of alternative systems of land-use.

The cost/benefit ratios prepared from the same data support the conclusion that the greatest relative benefits can be expected to accrue to the most intensive form of development.

Contribution to gross domestic product

One of the important considerations to be taken into account in assessing alternative development opportunities is the contribution which each type of development would, in the longer term, make towards the gross domestic product. After the point at which annual production becomes stabilised, the value of gross annual production from each type of development is estimated as shown in Table 11.5.

This illustrates that even under modest crop-yield assumptions, after full development is reached, the contribution to GDP resulting from agricultural development is some 2·7 times greater than that of plantations and 6·3 times that of intensified forest management.

Employment

Precise figures are not available to illustrate the relative labour demand

TABLE 11.5 Nepal: Terai – Estimates of gross production for possible development alternatives

Type of development	Gross annual production (Rupees per ha)	Gross production on 66 000 ha (million Rupees)
Intensified forest management	224	14·8
Plantations (55 t/ha)	525	34·6
Agriculture	1 420	93·7

created by each form of potential land utilisation at each stage in development. When a point of stability is reached, however, it is expected that the numbers of people shown in Table 11.6 should be directly and regularly employed in each type of development.

TABLE 11.6 Nepal: Terai – Employment effects of various possible development alternatives

Type of development	Number employed per 1 000 ha (minimum)	(maximum)	(mean)
Intensified forest management	5	10	7·5
Forest plantation	50	100	75
Agriculture (dry land)	750	1 500	1 125

Given the approximate magnitudes shown in Table 11.6, it is clear that agricultural development, even when partially mechanised, can provide significantly greater numbers of employment opportunities per unit area than any other form of development. Equally important, gross discounted costs per incremental employment opportunity would be relatively low for agricultural employment as is shown in Table 11.7.

TABLE 11.7 Nepal: Terai – Discounted cost per job created under various possible development alternatives

Type of development	Gross discounted costs (at 10%) per 1 000 ha	Incremental number employed per 1 000 ha	Discounted cost per job (Rupees)
Intensified forest management	388	5	77 600
Forest plantation	2 443	72·5	33 690
Agriculture	5 460	1 122·5	4 850

While net returns per worker would be higher in heavily mechanised forest plantations, levels of *per capita* income would be adequate by Nepalese standards for each course of development.

Secondary employment opportunities, particularly for processing and distribution, are expected to be higher for development based on agriculture than for either form of forest-based development.

Social aspects

Environmental considerations

The development of agriculture and agriculturally related enterprises can take its toll on the environment. Some developing countries feel that pollution of the environment is an inevitable cost of growth. For example, in one West African country, interested in forest industries development, the Minister of Agriculture expressed the view: 'It is better to have industry with pollution, than no industry and no pollution. Therefore, bring us the industries.'

This is not to imply that developing countries are not becoming increasingly aware of the environmental problems. At the global level, the problem is being recognised. The first United Nations Conference on

Human Environment in Stockholm in 1972 resulted in an action plan and the establishment of a fund to finance approved environmental plans. As a result, an agency, the UN Environment Programme was located in Nairobi. The World Bank found it necessary to establish an office of Environmental Affairs with specific responsibility for reviewing the environmental effects of the projects it considers for financing.

At the country and project level, environmental considerations refer to the possible impact of the project on the ecology, e.g. rainfall patterns, river flows, the scenic and aesthetic aspects; health effects on the population. There is also the impact on the rate of depletion of resources, caused by soil erosion, over-exploitation, overgrazing, by extraction at rates exceeding the maximum sustainable yield, e.g. in forestry and fishery ventures.

The erosion problem is usually the most insidious in agricultural development programmes. In St Lucia, for example, the banana industry grew very rapidly under uncontrolled conditions. The forest cover gave way to banana farms in areas which, because of their slopes and proneness to erosion, should never have been used for banana cultivation, particularly in the absence of a well-planned anti-erosion control programme. The country now has to pay for this, in terms of sharply reduced levels of soil fertility and flash floods resulting from changes in the ecosystem.

In a forest-utilisation project in Turkey's Antalya area, the main problem was one of location of a large pulp and paper plant, for which proximity to a reliable water supply is a major consideration. The ideal site from an economic point of view conflicted with plans for tourism. On the one hand, economic interests were concerned with existing road networks, and effluent treatment and disposal considerations; on the other, preservation of the area's natural beauty and historical sites was of first importance. Further work on selection of a site among alternative possibilities had to be undertaken and planning of the process within the plant had to take account of environmental safeguards.

In the Terai of Nepal, uncontrolled spontaneous settlements have resulted in wanton destruction of the forest resources, with only small quantities of timber retained by settlers for their dwellings. Some 70 per cent of the forest-covered land contains large tracts with low levels of soil fertility, incapable of supporting sustained agricultural production, which are prone also to erosion and flood damage. Under a resettlement project, arrangements would be made for the orderly use of forest resources and for the permanent maintenance of land unsuitable for agriculture under forest cover.

The experience with transmigration projects in Indonesia is especially relevant. The government's programmes aim at relieving the population pressures in Java and Bali by the transfer and settlement of transmigrant families to frontier areas mainly in Sumatra, Kalimantan and Sulawesi. The scale of the effort is shown by the fact that government, under its third five-year development plan (*Repelita III*, 1979–84) aims at the settlement of 500 000 families, mostly landless farm labourers and small part-time farmers. A recently approved World Bank loan and credit would resettle about 30 000 families as part of the plan.

A major element in the project's investments would involve the clearance of about 37 500 ha of forested land for settling the families, in addition to 7 500 ha for roads and public facilities. Thus the project would clear 1·25 ha per family of the food cropping land, of the total of 3·5 ha that each transmigrant family would receive, with the settlers developing the remaining 2·25 ha themselves, mainly with perennial tree crops.

The environmental considerations in such large-scale land clearing operations are important, having regard to the risks arising from possible soil erosion. Based on the nature of the soils, the question was whether sustained food crop cultivation is feasible on the marginal soils that characterise the settlement areas. Evidence exists that with proper cultivation, and application of farm inputs, the risks to sustained cultivation can be overcome.

The most appropriate methods of land clearing for opening new lands without damaging the soil resources, nor the valuable timber resources, had to be determined. Thus the merits of mechanical versus manual methods had to be assessed. In practice, a combination of both methods would be employed. The objective of the land-clearing operation, therefore, was to clear the vegetation with as little disturbance of the soil as possible so as to minimise subsequent erosion and to preclude the bringing to the surface of sub-soil with a low nutrient status. Strict criteria were laid down for the contractors doing the land clearing. Other preventive measures such as contour planting of grass strips, and encouraging farmers to plant food crops on the contour and trees along the contour lines were included in the project planning. Food crop production would be limited to land with 8 per cent slope or less to reduce the incidence of erosion, with tree crops limited to 8–15 per cent slopes, and steeper areas left in forest.

Fertiliser applications were planned to improve and maintain soil fertility. Sites already cleared would not be left exposed during the wet season. There would be a cover of crop stubble so as to build up organic matter. The use of a cover crop on the land, throughout the year, combined with contour planting of crops would reduce the risks arising from soil erosion.

All these measures were built into the project design, because in many tropical countries, the presence of a luscious green forest cover can readily give the erroneous impression of the presence of soils with high levels of fertility. However, it is quite common that the equilibrium between the soil-humus layer and forest cover is rapidly disturbed by the removal of the forest cover, with the resultant rapid losses of soil fertility arising from erosion due to heavy rainfall.

Apart from erosion, one project risk arising from the project's environmental effects would be the pressures arising from pests and diseases, as changes in the prevailing ecosystem occur. The project would make provision for this by getting the extension service to promote resistant crop strains, multiple varieties, crop rotation and intercropping.

The risk to settlers in their new environment from malaria would be tackled through a malaria-control programme associated with health-care packages and facilities. Intestinal parasitic diseases would be countered through provision of safe water supplies, waste disposal and initial treatment of settlers.

A further example of a major environmental consideration is the case of the Aswan High Dam in Egypt, which allowed perennial irrigation through stabilising the flows of the Nile. This has resulted in the raising of the water table and increasing soil salinisation, for which adequate drainage must be provided if these soils are to be made productive again.

The disease of schistosomiasis (bilharzia) affects millions of people in Egypt and perennial irrigation improved the habitat for the snail vector of the parasite disease. Under one World Bank financed project in Upper Egypt, for agricultural drainage, provisions were included for chemical control of the snails and treatment of affected individuals in line with recommendations of a specialised bilharzia-control expert.

The lesson to be learned from the foregoing examples is that for projects which include activities that disrupt the ecological balance, provision should be made, during the project-planning stages, to include the necessary and cost-effective safeguards, e.g. effluent-disposal systems in agricultural and forest products-processing industries, to counteract the negative environmental effects of activities financed by the project.

Multiplier effects

A project may have indirect effects such as multiplier effects and externalities which make themselves felt outside of the boundary of the project. Here we must differentiate between the project, i.e. investment and project actions, as opposed to the physical boundaries implicit in the concept of the project area. While economic analysis does not, in practice, measure these indirect benefits, largely because of the complexities of measurement, they can, sometimes be important in the overall economic growth process; the project planner should at least attempt to identify them and to gauge their importance at least broadly. For example, the project may reduce the rates of migration from rural to urban areas and thereby reduce the costs of urbanisation.

The project may also have an impact outside of the country's borders. This applies in particular to estate plantation crops, predominantly export-oriented, and where the bulk of international trade is accounted for by a few countries. External price effects, that would be induced by large planting schemes, could have harmful repercussions on other exporting countries.

Social effects of projects

It is sometimes argued that some agricultural projects can and do introduce negative social effects by further widening of the income disparities of beneficiaries that existed at the beginning of the project in the project area. Studies supporting this view are extremely limited.

However, one such study[7] of the Muda River Irrigation Project in Malaysia, with the aim of double-cropping of rice came up with the following conclusions: the Muda project implemented between 1966 and 1974, was designed to provide irrigation water for the double-cropping of rice on about 97 000 ha. The increases in income at the farm level were not evenly distributed across the already unequal patterns of income-distribution and therefore served to worsen that distribution.

The derivation of Gini Coefficients (an aggregate numerical measure of income inequality ranging from zero (perfect equality) to one (perfect inequality) for distribution of gross annual income (Gini = 0·406) and farm-size distribution (Gini = 0·360) shows that there are other factors besides farm size and/or conditions of land tenure which determine the distribution between the various classes of beneficiaries of the projects' benefits. Patterns of expenditure were shown to change with rising incomes, and these changes brought about improved nutrition and higher standards of material comfort. Thus a progressive reduction of the incidence of nutritional diseases occurred with the increasing introduction of double-cropping of rice. Evidence showed that the new dietary improvements were sufficient to prevent new cases of nutritional diseases from arising.

One interesting feature relates to the impact of the project on the labour market. Increased on-farm employment of family labour induced by double-cropping reduced the farm family's participation in the non-farm labour market and thus created job opportunities to be filled by others, and exerted an upward pressure on rural wage scales to the benefit of those with micro-farm units, and the landless.

The argument goes that in irrigation projects, especially the larger ones and in such projects involving installation of tubewells, that it is the rich landholding classes who monopolise the assets created by the project, making thereby the fullest use of the rural power structure. Those who first receive the tubewells can sell water at their price to their poorer neighbours. A similar situation arises with say, mechanisation projects, where the larger landholders sell tractor services to the smaller landholders. It is further claimed that, in the Indian subcontinent for example, these larger farmer-classes have strong political linkages, which they exploit fully when development projects are located in their areas.

The analysis of income impact by farm size of a rural-development project in one Near East country showed the following results: for a farm model of 0·8 ha the increase in net *per capita* income with the project would be 54 per cent; for the 2·0-ha model, 110 per cent; for the 3·0-ha model, 136 per cent; and for the 4·0-ha model, 85 per cent. Thus, the bulk of the beneficiaries would be the small farmers who would benefit most from the project. Thus the project would have a positive income-distribution effect, despite the fact that the sizeable increases for those with 0·8-ha model farms would still lie below the absolute poverty line determined for the country.

Project design and choice of components can do much to offset the adverse social impacts of projects. For example, the deliberate siting of a project in areas with large percentages of small farmers can do much to redress the situation. In land-settlement projects, priority can be given in the criteria for settler selection to the landless and/or those with uneconomic farming units. In irrigation projects, project design can provide water-delivery systems that ensure that the smallest-farm owners receive a reliable water supply. The system of water allocation can be so worked out as to ensure that the smallest farmers receive water supplies to guarantee their full cropping needs. Discriminatory water-recovery charges in favour of the smaller farmers could be applied.

Labour-intensive construction methods can be employed so as to increase employment opportunities for the tenant farmers and for the landless. Then, there is the distinct prospect of directing the projects' supporting services in favour of the small farmers. Thus, agricultural credit, extension and use of farm equipment are useful tools that could be heavily weighted in favour of the small farmers.

Social considerations in project formulation must start with an explicit recognition and definition of the target groups, with a view to improving the incomes, nutrition and overall quality of life. Agricultural credit projects and rural development projects represent two powerful approaches. The International Fund for Agricultural Development has actively pursued this in its lending operations. In a recent loan to Peru, the target group comprises the poorest farmers, living under hardship conditions, who have so far been overlooked by the country's lending institutions. In Pakistan, a recently approved loan, co-financed by IFAD with the World Bank, requires the borrowing institution to progressively increase its share of lending to small farmers over a stated time period. Furthermore, special provisions were made whereby the landless workers, previously ineligible because of collateral requirements could qualify for loans for approved expenditures.

If such credit approaches are to be successful, there are three indispensable requirements. The first is that the attitudes and approaches of the management and field staff of the national lending institutions must change. They must gear themselves through staff-training programmes, and the development of improved special procedures, to cater to the special requirements of small farmers. Second, technological packages appropriate to small farmers must be developed.

For example, farm machinery appropriate for large farmers may not answer the requirements of small farmers. In Thailand, the credit authorities feel that the absence of a technological package for rainfed agriculture is a severe constraint on lending to farmers with small farms. Thus, a demonstrated potential for remunerative production increases is the first guarantee for loan recovery. The third requirement is the organisation of the target group farmers at the local level to receive and productively employ such credits. Their active involvement in the decision-making process about the objectives and activities to be financed out of loan funds is critical.

While social considerations can be incorporated in a single project, they should logically be part of the country's overall development strategy. Nepal is a case in point. Very briefly, an IFAD Special Programming Mission to Nepal examined the country's overall development strategy with special reference to poverty orientation and proposed a pipeline of specific projects to support the recommendations for improvement.

Nepal has one of the lowest levels of *per capita* income in the world, with an extremely high share of the population falling below the minimum poverty line and very high levels of undernutrition. Three distinct geographic regions prevail – the Terai (lowlands), hills, and mountains. About two-thirds of the population live in the hills. A development strategy for Nepal must take account of these regional differences. In short, the strategy proposed called for intense utilisation of the abundant water resources of the Terai rapidly to boost up agricultural and food production, and make it the food granary of Nepal. The hill areas would intensify their production of horticultural crops and livestock with grain production limited to specially adapted areas.

While the proposed strategy focused on food production as the foundation for the development effort, it explicitly recognised that it is growth in employment which is primary. This leads to a growth in the income of low-income people and this provides the basis for their improved food uptake and nutrition. In turn, this generates the effective demand for increased food production. In short, the strategy recognised that attention must be given to both supply and demand factors, if Nepal is to meet its food needs. Supply factors would be dealt with through production-oriented investment and policy, while demand would be stimulated through increased employment and incomes for low-income people.

The package of investment proposals emerging from this strategy called for irrigation development in the Terai; small-scale irrigation development and horticultural development in the hills; a cottage and small-scale rural industries development to raise employment and income; and a small-farmer development project, discussed in more detail below.

It is to the credit of the government of Nepal that it has recognised that special efforts are required to promote the participation of the disadvantaged groups in rural development. The prevailing village administrative structure – called *panchayat* – did not permit this. Since the poorest populations in the villages lack social power, their voices are seldom heard in the village level assembly. They, therefore, cannot tap resources flowing in from outside, and their lack of credit-worthiness prevents them from qualifying for credit which is so basic to the production process. Thus in 1969–71, only nine per cent of the institutional credit reached small farmers.

To overcome this situation and ensure participation of the rural poor in development, it was imperative that they be organised to have a stronger voice collectively in the affairs of the community, and to pool their resources together as a group. This required, in turn, that their savings be mobilised and that the credit problem be solved.

A Small-Farmer Development Programme was thus established on a pilot basis in 1975. The programme sought to motivate the target group – small farmers and landless workers – to form small organisations of their own below the level of cooperatives, with a membership of about 15 to 20, so as to undertake a common income-raising activity based on group work-plans and group action. This was supported by an integrated programme and supervised credit, extension and technical back-up. The Agricultural Development Bank of Nepal (ADBN) implements the programme in the field with the assistance of various line agencies.

The groups were to receive credit for individual members as well as for the group as a whole on the basis of group demand under group liability with credible income-raising action plans only, and without any tangible collateral. Production enterprises in crop, livestock, horticulture and cottage industries were emphasised for loans from the ADBN. The groups were required to promote their own group saving-fund from member contributions to provide consumption, distress and emergency loans to the members.

By early 1979, 94 groups comprising 1 265 farm families had been established in the pilot areas. In the Nuwakot pilot area (40 groups), total credits amounted to 1·33 million rupees (about US$110 000) at an average rate of Rupees 1 770 per family. Livestock activities dominated, with milk, buffaloes, goats and sheep accounting for 83 per cent; crop production, 11 per cent and others 6 per cent. The group's own savings totalled Rupees 50 000. Repayment on the whole was good and on schedule.

The success demonstrated by these two pilot areas induced the government to undertake an expanded programme at a scale that could be considered for assistance from external financiers. One determinant of the rate at which further growth on a sound basis can occur, is the availability of sensitive and capable group organisers, for which adequate training provision would have to be made.

The foregoing demonstrates a working model for participatory rural development, that could be applied to many other developing countries. The essence of this is getting the target groups to share actively in decisions and actions related to improving the quality of their lives. This further illustrates the role that rural credit projects can play in resolving the problem of income disparities within micro-regions.

The role of rural-development projects as an instrument for bringing social considerations into the development process also requires attention. The very definition of a rural-development project, as one which delivers at least 50 per cent of its benefits to those falling below the country's poverty line, indicates the social orientation of such projects.

The social infrastructure components, such as potable water, health facilities and schools clearly generate social benefits. Improved water supplies and availability are highly welcome by the female population, by saving time and effort, which can be devoted to other activities, e.g. home gardening and dressmaking. Improved health facilities, particularly involving low-cost delivery systems, reduce the incidence of family illness for which family debt may have to be incurred. With regard to improved education facilities, many rural societies attach high priority to this as an important element for their children to move up the social ladder.

In fact, the tendency of many countries to concentrate their social investments in urban areas is one factor accounting for the urban drift from rural areas. Thus a positive benefit from social investments in rural areas is a reduction of such migration rates.

It will be recalled, however, that production investment components invariably dominate the investment-mix in rural development projects. A brief review of the impact of the rural development programmes in Mexico (PIDER), discussed in Chapter 4, throws much light on the special role that social considerations must play during the concept-definition and formulation phases of a project, if such programmes are to deliver the planned levels of benefits to the target beneficiaries.

Mexico's rural-development programmes are monitored by the Research Centre for Rural Development (CIDER), and their findings formed the basis for a World Bank Staff working paper on the subject.

The following discussion focuses mainly on the *ejidatarios*, small farmers, who are members of the *ejido* cooperative groups with joint ownership of land.

A major finding was that PIDER investments did not generate the expected increases in production, income and employment, largely due to inadequate treatment of social factors. In fact, the existing social orientation within the *ejidos* combined with the communal uses of pasture land, resulted in the investments favouring the wealthier *ejidatarios* to the disadvantage of the poorer ones, and social conflicts and clashes arose. Since the cattle-holding peasants control political power in the *ejidos*, attempts at pasture-regulation were not supported by the power groups. Thus equitable use of the common pastures and proportional payment for use of the resource could not be realised.

In order to redress the situation, it was recommended that the need for building the social organisation for cattle development should precede, or at least directly accompany, the building up of the physical infrastructure for livestock units. Social organisation requires the engineering of social arrangements, establishing patterns of work cooperation and distribution of benefits, assigning responsibilities within the involved communities and ensuring linkages with the other village activities.

On the question of people's participation, it was suggested that the need exists strongly to formalise the village-level organisational structure and to institutionalise the process of people's participation. This is imperative if village-level investments are to be maintained and mechanisms for programme-coordination developed. Thus, the villagers themselves, rather than just the village *ejido* leaders, would be involved. This could contribute substantially to the promotion of self-sustained development.

One very interesting finding of the CIDER evaluation study of the PIDER rural development programmes is the need for improved research methodology in analysing the impact of investments at the level of the ultimate beneficiary. It is argued that the fundamental unit in the peasant agricultural sector is not the farm enterprise, which strictly represents an economic unit of production, but is instead the family household which represents both a unit of production and a unit of consumption. There is the need to know more about how the household decision-making process operates with regard to the production and consumption issues. How do these influences, for example, affect changes in cropping patterns?

Both the foregoing cases in Nepal and Mexico bring out the following major points. First, that at the village level, income disparities prevail and these are associated directly with the social and power structure. Experience in Indonesia supports this view, and within what may be called a 'poverty region', the income differences within groups in a village may be wider than between villages. Second, the poverty groups within the villages lack the cohesion to be able to tap effectively external resource flows reaching the villages. Therefore, the answer to the problem is to organise themselves so as to receive and direct the use of these resource flows.

While this possibility of group organisation provides a plausible and viable option for millions of the disadvantaged rural poor to improve their living conditions, it is not devoid of risk. Implicit in all this is the possibility of relative changes in the power structure, with potential political implications. Clearly, therefore, an indispensible requirement for success for such new approaches to development, is a greater social consciousness on the part of governments and national planning agencies, and a national determination to improve the quality of life of their poor rural populations.

References

1 Leonard Joy and Philip Payne, *Food and Nutrition Planning*, Consultants' Report Series, No. 35, ESN:CRS/75/34 FAO, Rome, 1975, p. 63.

2 P. Dasgupta, S. A. Marglin and A. K. Sen, *Guidelines for Project Evaluation,* UN/UNIDO, New York.

3 John R. Hansen, *A Guide to the UNIDO Guidelines: Social Benefit/Cost Analysis in Developing Countries*, UNIDO, 1975, ID 75–6769, (unpublished).

4 Ian M. Little and James A. Mirrlees, *Manual of Industrial Project Analysis in Developing Countries*, Vol. II, *Social Cost/Benefit Ananlysis*, Development Centre of the Organisation for Economic Cooperation and Development, 1969.

5 Lyn Squire and Herman G. van der Tak, *Economic Analysis of Projects*, A World Bank Research Publication, Johns Hopkins University Press, Baltimore and London, 1975.

6 J. Price Gittinger, *Economic Analysis of Agricultural Projects*, The Economic Development Institute, International Bank for Reconstruction and Development, Johns Hopkins University Press, Baltimore, 1972, Chapters 1 and 2.

7 FAO/World Bank Cooperative Programme, *The Muda Study*, Vols I and II, FAO, Rome, 1975.

PART IV

Case studies

12 Nigeria – Funtua agricultural development project

Introduction and background

While Nigeria is the only African country belonging to the OPEC countries, it still faces considerable problems in its economic development. This arises because of the fact that about three-quarters of Nigeria's population, which stood at about 75 million in 1975, depends on agriculture, which acccounts for about 28 per cent of GDP, 5 per cent of total export earnings and 75 per cent of non-oil exports. Small land-holders with low levels of technology dominate agricultural production, and are the main focus of the country's rural poverty. Thus, while in 1975, average annual *per capita* income was estimated at US$327, the average for the rural areas was only US$60. These urban–rural income disparities need to be redressed, if the bulk of Nigeria's population is to enjoy the country's overall economic growth.

While this case study is located in Africa's richest and most populous country, it has strong relevance for many countries in the African continent. This is so because the case study focuses on low-income beneficiaries. It illustrates the potential for growth under conditions of rainfed agriculture, and the means by which technology can be introduced into a static agriculture, involving large numbers of small farmers and achieve sizeable increases in farm income. The crops covered in this case study include both cash crops and staple food crops which are typical in many other African countries. The strategy pursued in this particular case is widely replicable because it seeks to provide a concentration of support services to areas of reasonable potential and dense population.

Nigeria's economic growth has been increasingly dependent on oil. In 1974, oil accounted for 41 per cent of GDP, 93 per cent of export earnings and 82 per cent of total government revenues. In the period 1972–75, the GDP grew at an average rate of 6·5 per cent per year. In contrast, the growth of the agricultural sector has averaged only 1·3 per cent per annum, from 1970 to 75, or about one-half of the average annual population growth rate. With rising population and incomes, domestic food production cannot keep pace with demand for food. Traditional export crops declined from ₦309[1] million to ₦265 million, while total exports including petroleum rose from ₦852 million to ₦4 848 million in the same period. Imports of food and live animals in the same period increased five-fold from ₦60 million to about ₦300 million.

The government of Nigeria fully recognises the need to reverse this trend, and to diversify the economy by investing its substantial revenues from oil into other economic sectors, including agriculture. The third five-year development plan (1975–80) earmarks a total of ₦2·2 billion (US$3·56 billion) for the agricultural sector which aims at a growth rate of 5 per cent. The programme proposed for achieving this growth rate would involve irrigation development of several major river basins; the spread of modern farming techniques to smallholders; and improved distribution of inputs and support for various integrated development programmes. Rural development would be supported with improved road networks, rural electrification, health and social services.

In the short run, government aims to improve internal price stability in basic foodcrops, and to ensure food supplies for the rapidly expanding urban centres. In the long run the aim is to make the country's agriculture efficient and thereby reduce migration to urban areas and to expand its agricultural exports.

The Funtua project, under review, represents one of a series of projects, that the government, with World Bank assistance, has undertaken to achieve these ends. Three integrated agricultural-development projects undertaken in 1974 included projects in Funtua, Gusau and Gombe, for which World Bank loans, totalling US$69 million, were obtained. While these projects have all experienced cost overruns, arising largely from rapid, internal inflation, their initial performance has been sufficiently promising to enable government to pursue similar projects. Two such projects, Ayangba and Lafia received World Bank assistance in 1977, totalling US$62 million. In addition to these five agricultural-development projects, other major projects, undertaken with World Bank assistance since 1971, include two cocoa projects, one rice project, one livestock project, and three oil palm projects, involving Bank loans of US$131·2 million.

These total Bank loans of about US$262 million to Nigeria during the period 1971–7 demonstrate clearly the importance of the role that agricultural investment projects play in Nigeria's approach to solving its agricultural developmental problems.

The project area

Location The project area covers some 7 500 km^2 or about 10 per cent of the total area of North Central State (NCS). It comprises the whole of the Funtua Administrative Division of Katsina Province in the northern half of NCS, which includes the five southernmost districts of the Katsina emirate.

Climate Annual rainfall, confined almost entirely to the six months, April through September, is of the order of 1 000 mm with a 75 per cent probability of obtaining 1 000 mm. Between October and March, there is virtually no rain; this long, dry season is rendered very harsh by hot dry north-easterly harmattan winds blowing from the Sahara.

Beginning in April, when there is a 50 per cent chance of obtaining 35 mm of rain through September, which has a 50 per cent probability of more than 215 mm, rainfall distribution is well suited to a range of annual crops. However, due to limitations in timely land preparation and to a labour bottleneck of planting time, few farmers can plant all their crops at the optimum time. Food crops are given priority over cash crops, and the success of the latter is dependent on the farmer's residual input resources, and because they are planted late, on the vagaries of climate; an early end to the rainy season often vitiates most of the work put into these crops.

Water There are few perennial streams in the area, and groundwater supplies are severely limited by the presence of the impervious underlying basement complex. With increasing pressure on land, the flow period of streams has diminished markedly. Wells commonly fail during the dry months, leading to prolonged water shortages which are becoming an increasingly critical issue for both human and livestock populations. Fortunately, topography is such that construction of small reservoirs appears a practicable solution to the water problem.

Topography and soils The area is underlain by rocks of the basement complex, which is predominantly granite. The landform is gently undulating at an altitude of 600 to 900 m with occasional granite inselbergs. Many small, ephemeral streams occur throughout the area.

Leached ferruginous tropical soils are common with well-developed sub-soil horizons containing some 30 per cent clay. Drainage ranges from good to rather poor. In the southern half of the area, silt content is high which leads to surface capping.

Soils are weakly acid, with pH in the range 5·5 to 6·5, and are deficient in nitrogen and available phosphate, particularly after prolonged cropping. Attempts to maintain the fertility of farms close to settlements is made by using animal manure at between two to three tonnes per ha. Recent investigations have demonstrated an apparent boron deficiency and its solution, by the use of boronated superphosphate. Fertiliser responses by most major crops have been established with some precision at low application rates. Field trials involving higher levels of application, mainly for cereals such as maize and sorghum, are presently underway in the project area.

Vegetation Natural vegetation is savannah woodland typified by *Isoberlinia doka* in association with grasses mainly of *Andropogon* and *Hyparrhenia* genera. Over much of the area, the savannah has been replaced, following clearing and farming, by parkland dominated by such trees as *Parkia clappertoniana* and *Adansonia digitata* (baobab). These trees are protected by law and provide valuable dietary supplements from seeds and leaves. In wetter areas, which have been heavily farmed, spear grass (*Imperata cylindrica*) is a troublesome weed. Additionally, the parasitic weed *Striga hemontheca* causes significant yield-reductions in sorghum.

Population The project area is occupied by the Hausa people, most of whom are devout Muslims. Literacy in English, Nigeria's official language, and Hausa, the vernacular, is low among adults in rural areas, but is rising rapidly with the younger generation. Available data are scant. The last census, from which data are available was carried out in 1973 and remains the best source of population data until further census is carried out. Such data will be of immeasurable value in detailed project planning and evaluation. The only other possible source available at present, local authority returns on numbers of tax payers, whilst collected annually, vary in accuracy in accordance with the diligence of the different officials charged with tax-collection responsibilities in the districts. Information from such sources indicates only 60–80 per cent of the adult male population that has been imputed from census data. According to the census, population of the five districts constituting the project area totalled 698 748 in 1963. Using a growth rate of 2·5 per cent annually, population of the project area currently should be about 905 000.

Data on numbers of persons per family in the project area are also sparse. No demographic surveys have been carried out. However, indications from tax data, point to an average of 4·22. Local opinion does not confirm this, suggesting that family size approximates 6·5 individuals. This figure in turn is lower than that obtained in a detailed survey of three villages in Zaria Province, some 60 km south of the project area. The latter arrived at a mean figure of 8·5 persons per family. This is substantially higher than the figure suggested by the Federal Office of Statistics for Zaria Province as a whole. Because of the spread between family size derived from tax data on the one hand, and a micro-sample in a neighbouring province on the other, 6·5 was arbitrarily selected as representing average family size. This was in accordance with local opinion and falls between the extremes otherwise suggested. Thus, on the basis of a family size of 6·5, and a population of 905 000, there are now estimated to be some 139 230 families in the project area.

Numbers of farming families were calculated on the assumption that about 28 per cent would be engaged primarily in non-agricultural activities. This figure was derived from census data and observations on non-agricultural activities in the area. On this basis, it is currently estimated that there are about 100 000 farming families in the project area.

The population is divided between five districts of Funtua Division and resides in 77 village areas. The estimated situation in 1974 was as shown in Table 12.1.

TABLE 12.1 Nigeria: Funtua Division: Population estimates by district

District	Village areas	Population
Funtua	16	229 000
Bakori	19	233 000
Malumfashi	22	228 000
Kankara	12	132 500
Faskari	8	82 500
Total	77	905 000

The social structure of the area is significant because of the relative strength and influence of the hierarchical structure of the local authority. The project comes within the jurisdiction of the Emir of Katsina whose hereditary authority is delegated through salaried district heads and village heads.

Among their other functions, village heads are responsible for village-level administration of land-tenure arrangements. The system is based on Islamic law and on customary usufructory rights. In theory, it provides for the allocation of such rights to land which is not yet so allocated, to any native Nigerian who expresses the intention of settling and using land within the village area. In practice, most land has been allocated to families, the heads of which sub-allocate to family members in accordance with assessed need and capability.

As representatives of the local authority, village heads have considerable political influence within their areas. Although this influence is being eroded in urban areas, such as the wards of Funtua town, it is still considerable in the villages. For this reason, every effort would be made to obtain the support of village heads for the project by acquainting them with proposed and on-going developments within their areas. District heads would be similarly consulted and informed. Furthermore, the project makes provision for local authority representation on the project committee.

Farm size Information on farm size is scant. It is possible to impute a possible mean farm size on the basis of assumed population, family size and estimated area of cultivated land. The latter figure is estimated by NCS at 390 000 ha or 52 per cent of total project area. Given 100 000 farm families, then average size of cultivated holding, i.e. excluding fallow, would be 3·9 ha. This figure is in reasonable accord with mean farm size established by a detailed survey in a similar area some 60 km to the south.

There are some very large farms within the area. These farms would normally receive credit through commercial banks; the project would help owners of large farms by employing a farm management specialist to advise them.

Communications The project area is well served by main roads. Additionally, the Nigerian Railway Corporation has a spur line from its main northern route, that runs from Zaria via Funtua to Kaura Namoda in North West State. A major road runs parallel with the line through the south and west of the project area.

Unfortunately, the basic communication network barely extends beyond that described above. Minor roads are non-existent or in poor repair. Communities are seasonally cut off from each other and the outside world. Transportation of produce and inputs to rural areas is both difficult and expensive. Attempts to improve the situation by self-help activities and by the local authority works department are hampered by shortages of funds and executive capacity.

Farm and crop development

The major crops in the project area are sorghum and millet, cotton, groundnuts and cowpeas. Minor crops include maize, sugar cane, chillies and okra. Production data for cotton and groundnuts are compiled from purchases by the Northern States Marketing Board (NSMB). No production data are available for other crops; nor is there other than qualitative information on farmers' yields, farm sizes or cropping patterns. However, a detailed economic survey of three villages within 60 km of the project area has been carried out and published recently. Ecological and sociological conditions are comparable in many respects between the two areas and, for this reason, the survey has been used as a basis for the formulation of assumptions on present and projected farming activities.

Cattle, sheep and goats are kept throughout the project area by farmers, and in substantially larger numbers by semi-nomadic pastoralists (Fulani). The latter utilise fallow land (a rapidly diminishing resource), forest and grazing reserves, and crop residues for their livestock. They exist in an increasingly uneasy symbiosis with settled cultivators as pressure on land increases with rising populations. Livestock-management systems are generally poor and productivity is low. There are no reliable data on stock numbers. Cultivations are carried out mainly by hand-hoe and cutlass. However, some 7 000 farmers (7 per cent) are reported to own work-oxen and equipment. In 1972/73 some 150 ha were ploughed and some 400 ha were ridged mechanically, using tractor-hiring services.

From May to October, the widest range between mean monthly maximum and minimum temperatures is 38 °C to 21 °C. At project area altitudes of 600 to 900 m, cold spells do not cause any problems and rainfall is the major factor limiting crop growth. Monthly and annual data for two stations were examined. Instead of presenting these in the traditional manner, they have been arranged in rank order to provide an estimate of probability. Thus the modal value (half above and half below) provided a demonstrable 50 per cent probability not normally associated with arithmetic averages. However, for a basically subsistence agriculture a more stringent requirement is appropriate and an 80 per cent probability was chosen. At this probability rainfall is still quite adequate for crop growth, and even in April, one can expect 20 mm. Rainfall distribution is critical to favourable production, and the bumper crop years have always been associated with seasons with extended rainfall at the beginning and end of the season. Basically the farmer is a pretty busy fellow who has limited resources for carrying out his land preparation and subsequent cultivation. Thus eight years out of ten he can expect sufficient rain in April to allow timely cultivations and subsequent planting operations. If the sequence is broken then late planting particularly of cotton and sorghum occurs which results in low yieids. Thus, provision of ox-drawn equipment should allow for more timely land preparation. At the end of the season, sometimes in September, and certainly in October, it is important to conserve soil moisture and reduce water run-off. Thus water-conservation measures such as contour bunding and the development of better ground cover (such as closer planting of groundnuts) are all techniques that should be encouraged. A progressive farmer, if taking such measures, should have little difficulty in improving his level of output.

Farming systems Although no detailed surveys have been carried out, field observations tend to support the findings of one study which indicate a mean farm size of 3·9 ha. This is somewhat larger than the size originally estimated by staff of the Ministry of Agriculture and Natural Resources (MANR), which is thought to be biased heavily towards the relatively densely farmed area near Funtua.

Cropping patterns

Intercropping is commonly practised in a wide variety of combinations. The cropping pattern of a typical farm, on the basis of adjusted areas of sole crops might be as shown in Table 12.2

TABLE 12.2 Nigeria: Funtua – Cropping pattern on typical farm

Crop	**Area planted** (ha)
Sorghum	1·20
Millet and other cereals	0·50
Cotton	0·90
Groundnuts and cowpeas	0·45
Others and fallow	0·85
Total	3·90

This pattern might be expected to change as a result of the project which would promote sole cropping and the introduction of improved production practices for cotton, sorghum, maize, groundnuts and cowpeas, and would result in changes in areas planted to different crops. Overall farm-size would not be affected by the project, however, the amount of land in fallow each year is likely to decrease. This decrease in fallow is currently resulting in marked decline in soil fertility, particularly in areas furthest from the homesteads which receive no farmyard manure. Some of these deficiencies would be made up in the short term by supply of fertilisers, and in the long term through the introduction of pasture. A summary of changes in cropping patterns that the project would introduce are shown in Table 12.3. For each major crop, the project's targets for introduction of improved practices versus traditional practices is shown, in respect of number of farmers and gross area.

Project description and detailed features

The project, as worked out at appraisal, would be carried out over a five-year investment period, 1975/76–79/80, and would involve the following activities:

1 Constructing about 1 500 km of low-cost agricultural roads, 85 earth dams, 160 *tapkis* (ponds), and associated soil conservation structures.
2 Constructing an administrative centre in Funtua, five development unit centres, and 77 farm service centres.
3 Rehabilitating Daudawa seed-multiplication farm and the expanding and equipping of an additional farm at Malumfashi.
4 Provision of adequate farmer extension, and credit and marketing services, including staff, vehicles and equipment.
5 Provision of seasonal and medium-term loans to farmers for the purchase of crop inputs and farm equipment.
6 Establishing an evaluation unit to evaluate the project.
7 Preparation of plans for the continued provision of project services after the investment phase of the project is completed.

The project would be managed by the Funtua Project Management Unit (FPMU) which would carry out the day-to-day management operations. FPMU would be responsible to the Funtua Project Executive Committee (FPEC) for budgetary and policy control. A committee would be set up to coordinate project activities at federal level.

TABLE 12.3 Nigeria: Funtua agriculture development project – Summary of changes in cropping patterns

	Without-project			With-project		
Crop	**Total farmers**	**Gross area (ha)**	**Per cent of cropped area**	**Total farmers**	**Gross area (ha)**	**Per cent of cropped area**
Cotton						
Traditional	95 600	85 550	22	51 500	36 050	9
Improved	3 000	3 150	1	27 000	63 000	16
Sorghum						
Traditional	100 000	120 000	31	51 000	61 200	16
Improved	–	–	–	60 000	30 000	8
Maize						
Traditional	20 000	5 000	1	20 000	5 000	1
Improved	–	–	–	47 500	27 500	7
Groundnuts						
Traditional	50 000	32 500	9	25 500	12 750	3
Improved	–	–	–	30 000	22 500	6
Millet	100 000	55 000	14	111 000	55 000	14
Cowpeas	20 000	3 800	1	20 000	5 000	1
Fallow and others		85 000	21		72 000	19
		390 000	100		390 000	100

Detailed features

Agricultural road development The project would provide a network of 1 500 km of agricultural roads, of a design adequate to allow light traffic during most of the wet season, and heavy trucks during the dry season. The roads would permit the efficient evacuation of farm produce, the timely supply of inputs, the effective movement of project service staff, and much improved general access to farms and hamlets. This basic infrastructure would allow other government programmes including health and education to expand into areas which currently are inaccessible. In view of an acute shortage of local contracting capacity, and the probable lack of interest of foreign contractors in work of this sort, construction would be undertaken by force account. This would be implemented by a project operated and managed field-construction unit that would make some use of small contractors and would thereby assist in their development. Maintenance of roads and dams would be the responsibility of the Katsina Local Authority which has a works department. In order to build up KLA operational effectiveness, KLA road-maintenance staff would be seconded to the project for training and operational control. The project would provide, during the investment period, funds for the purchase and operation of the necessary maintenance equipment.

Water development The project would set up and operate a unit to construct some 85 small and medium-sized earth dams, each of a minimum capacity of 100 000 $m,^3$ equipped with simple systems to reticulate the water away from the reservoir. The reservoirs would supply water for human and livestock needs and would be located in strategic areas. They would also be used to supply water for compaction purposes during road construction. In addition, some 160 *tapkis* (ponds) would be developed to provide a better distribution of water for livestock for all but the last three months of the dry season. In a very few instances, the dams and *tapkis*, where land and water availability allows, would be used for the irrigation of village vegetable production. Abundant dam sites are available.

Soil conservation In association with the water-development programme, soil-conservation measures would be undertaken within the catchment areas of the dams (about 5 km^2 each). These would include the construction of cut-off ditches, artificial waterways and contour ridging and be aimed at reducing siltation of the dams and controlling erosion. Further soil-conservation measures would be undertaken in providing the farmers with marker ridges to assist in contour realignment of crop ridges. During the project investment period, some 2 700 km^2 or 36 per cent of the project area would be so protected; these programmes should be continued beyond the investment period.

Building development The project would include construction of houses, offices and stores needed for project services. It would include the construction of 350 houses of various sizes, a project office, five development centre offices, a railhead store and other facilities. In addition, 77 farm service centres, each including a 500-tonne capacity warehouse would be constructed, and improvements made to three market depots associated with existing and future cotton ginneries. Those houses and buildings not required by the Ministry of Agriculture and

Natural Resources (MANR) after the project investment period, would be taken over by other government departments operating within the area.

Seed-multiplication farms The project would make improvements to the existing seed multiplication farm at Daudawa; and expand and develop the Malumfashi seed farm. Investments would include farm buildings, water supplies, farm machinery and equipment. These seed farms would, in association with nearby large farmers, provide project and other state farmers with improved seeds, act as demonstration and research centres, and provide practical training in farm management.

Training facilities The project would expand boarding accommodation at the Daudawa training centre from 20 to 40 trainees and construct a new centre at Malumfashi, also catering for 40 residential trainees. Initially, the centres would be used for project staff training, but progressively emphasis would be on farmer training.

Farmer support services The project would expand and intensify support services for farmers. It would increase the farmer extension personnel ratio from the current 1:2 440 to 1:240. This would involve an increase in extension agents from 41 to 420. The extension personnel would include, besides basic extension agents, personnel specialised in livestock husbandry, seed multiplication, commercial farm management, farmer training, and farm equipment. The tractor-hire service now operated by MANR at a loss would be phased out, and gradually replaced by project sponsored and assisted private operators. A credit service would be established to operate a project controlled credit-revolving fund and the contingent services to farmers. A marketing service would also be established to improve market liaison, quality control, price information, produce weighing and to control traders. The credit and marketing services would employ some 260 staff of various grades; of these only about 30, mainly produce examiners, are now employed by MANR. Consequently, the majority of personnel employed in these services would receive in-service training from the project.

Other MANR services such as veterinary, fisheries and forestry would continue to work within the project area, and would coordinate their activities with FPMU.

Farm inputs Under the project, seasonal inputs would be made available to farmers for cash and credit. Such inputs and equipment are estimated to amount, during the project investment period, to 56 000 tonnes of fertiliser, 4 000 tonnes improved seeds, 2 000 tonnes insecticides, 47 000 ultra-low-volume insecticide sprayers, 10 000 ox carts, 10 000 ox ploughs and 100 tractors and associated implements. Inputs and equipment would be sold for cash or credit at the farm service centres.

Project monitoring and evaluation A project Monitoring and Evaluation Unit (PMEU) would be established by the Federal Ministry of Agriculture and Natural Resources (FMANR). PMEU would comprise two sections, the first would technically review and oversee financial records and statements in respect of the project, and the second would develop a system of evaluation and analysis to provide the Federal Military Government of Nigeria (FMG), state governments and other institutions, relevant economic and social data on the success or failure of project policies. Such information would be vital to long-term agricultural and rural planning in Nigeria. FMG agreed to employ consultants to staff the second section. PMEU would give technical direction to FPMU administered field-evaluation units to ensure adequate collection of information for PMEU analysis.

Rationale for choice of project components

Whilst this project is called an integrated agricultural-development project, it does not differ markedly in concept from a rural-development project. The reader will note that the components include provision of physical infrastructure, e.g. agricultural roads; directly productive components, such as agricultural credit and seed multiplication; and social components such as water for drinking, from the proposed water development.

The aggregation of these components represents an attempt to overcome the constraints facing the area's development through a set of coordinated actions. The rationale for inclusion of some of the more important components is discussed below.

Thus, the agricultural-roads component is justified on the basis that the project cannot succeed unless the constraint posed by roads is removed. Thus, an estimated 1 500 km of agricultural roads was required that would permit most farmers to be within 2 km of a road in the Funtua area. The road network – largely unsurfaced – would permit the delivery of farm inputs, transport of produce for marketing and improved services by the agricultural extension staff. Social benefits would also result including better communication between the villages. The average overall road-construction costs were estimated at ₦1 900 (US$2 900) per km. Apart from capital costs, the project provided budgetary funds for road maintenance.

The water-development and soil-conservation components were included because of the acute shortage of water supplies for domestic use during the dry season and the early part of the wet season. In the project area, the existing wells, dams and ponds were estimated to meet only 10–20 per cent of total demand requirements. At peak periods of shortages, water, head-carried or by donkeys, fetched prices of ₦0·10 (US$0.15) per 20 litres. The most appropriate technical solution to the problem was to exploit surface water supplies and impound wet season (June–September) run-off in small earth dams for which topographical maps suggested a network of small dams, of about 120 000 m^3 in total capacity each, should be built. The dam would be fed by or run-off from, a catchment area of about 3 to 4 km^2 in area and would meet the year-round demands of 2 000 people and their livestock.

Soil-conservation practices in the project area would be closely linked with water development because the most erosion-prone areas are precisely in the catchment areas of small streams and rivers. Soil erosion has its origin mainly in overstocking and concentration of livestock at the sparsely located watering points and through poor cultivation practices. Thus, dam catchments would require special conservation measures to prevent erosion and siltation.

The farm-inputs component envisages the provision of seasonal inputs and equipment to farmers for cash and credit. Fertilisers, insecticides, improved seed and light farm equipment, including ox carts and ploughs would be provided. Thus the project would procure, and distribute these items and decide on the suitability and composition of any package of inputs offered to farmers. The expected value of these inputs and equipment required by the project was estimated at ₦18·1 million, of which 75 per cent of the seasonal inputs and 100 per cent of the equipment would be sold on credit.

While the traditional agriculture in Funtua used almost no modern agricultural inputs, the project could not possibly achieve its goals without the availability and use of these inputs. Demand would have to be created, and farmers trained in the use of inputs and equipment. The project's first task was to ensure the ready availability of inputs and equipment since suppliers of farm inputs in the project were for practical purposes non-existent. Thus, the commercial activities of the project's management would be undertaken by a chief marketing officer, who would be responsible for the logistics of acquiring, moving and distributing these inputs and equipment. Fertiliser requirements by year five of the project would exceed 22 000 tonnes, and some 14 000 sprayer kits would be required in the same year.

These project actions called for the construction of a 4 000-tonne storage facility at Funtua railway yard, and 77 village stores each with a 500-tonne storage capacity, while private truckers would undertake the rail-to-village transport.

The existing tractor and equipment services provided by government in the project area were limited and uneconomic and would be replaced by contract tractor operations undertaken by private entrepreneurs who would be financed under the project. Existing transport contractors, small businessmen and large farmers would be the beneficiaries of about 100 loans for acquiring tractor units with attachments, and of the development of maintenance services with the aim of tractor cultivation of 6 000 to 8 000 ha annually by the fifth year of the project.

The provision of farm credit to enable farmers to acquire these inputs is indispensable to the project's success. The logic behind this is simple: the large majority of farmers in the project area were small farmers and received almost no commercial credit, since the commercial banks were just not equipped for loan operation to masses of small farmers. The latter did receive some credit from local traders and other private sources but such credit was unreliable, costly and not geared to meet the requirements of a productive agriculture. Previous efforts of cooperative societies failed for various reasons, including lack of ready availability of inputs and lack of technical supporting services.

The project, therefore, provided organisationally for a credit manager who would head the Credit Services Section of the project with the main task of administering the revolving-credit fund, established under the project. The estimated credit requirements annually through to year five, are shown in Table 12.4.

It was necessary to work out procedures for the operation of the credit fund. For example, each borrower's loan application would be approved by the hamlet or village head, or a farmer's association

TABLE 12.4 Nigeria: Funtua – Estimates of credit requirements

	Credit requirements (₦ '000)				
	Year 1	**Year 2**	**Year 3**	**Year 4**	**Year 5**
Fertilisers	127·8	319·2	740·9	1 221·7	1 601·5
Insecticides, seeds, battery sets	95·9	313·1	691·5	1 037·9	1 258·5
Sprayers, ox-drawn equipment and tractor equipment	86·0	222·1	511·9	715·7	904·6
Total	309·7	854·4	1 944·3	2 975·3	3 764·6

approved by the village authorities, as evidence of the applicant's credit-worthiness. These group social pressures should operate to reduce defaults on repayments by borrowers. Various lending terms for on-lending to farmers were worked out subject to experience gained during implementation. For example, seasonal loans for farm inputs would be repayable within 12 months at a flat service-charge rate of 10 per cent, and no down payments would be required.

Project monitoring and evaluation component

This would be established by the Federal Ministry of Agriculture and Natural Resources (FMANR). As noted, this would have two sections, one concerned with the technical review and financial records and statements related to the project and the other concerned with developing a system of evaluation and analysis of economic and social data on the success or failure of project policies.

The logic behind the establishment of a project monitoring and evaluation unit that would be established at Kadusa was that rural-development projects of the size and scope of the Funtua project can have far-reaching effects on the people in the project area and on the economy of the state and the nation. The unit would also monitor and evaluate the Gusau and Gombe Agricultural Projects in addition to the Funtua project.

The objectives of the unit would be as follows:

1 To guide project management.
2 To help establishment of project goals and scheduling of project activities.
3 To measure effectiveness and efficiency of project activities.
4 To inform state and national governments, and others of changes stemming from project activities.
5 To enhance knowledge of the rural-development process for future development planning.
6 To permit the continuous reappraisal of project goals and assumptions underlying project design, particularly those aspects affecting farm credit and input supply, and to allow for progressive modification of project objectives and structures.

Evaluation must include such standard economic criteria as changes in gross and net incomes of participants, changes in volume and value of production and measures of efficiency in use of project resources. It must include also such direct and indirect benefits as changes in efficiency and effectiveness of the marketing system, changes in employment, changes in the volume and type of business in the region, changes in tax revenues and expenditures, and changes in other economic and social indicators that would show direct and indirect results of project activities. Equally important may be changes in patterns of consumption, in participation in education, in health, and in other measures for the well being of people in the project area.

Precise types of measures to be made and specific methods of data collection and analyses would be developed by the evaluation staff on site. Some of the data needed would be generated by project records. Other data would be collected by surveys and from secondary sources outside the project area. The proposed investment of substantial expertise and manpower in developing and conducting surveys, and analyses designed to measure meaningfully the parameters of rural development, should generate considerable benefits, both for future planning in Nigeria and for the general knowledge of the rural-development process.

If the broader impact of the project is to be assessed, there is no alternative to undertaking baseline surveys before the effects of the project begin to manifest themselves, and to continue the assessment of changes brought about by project actions during and after the implementation period. No organisation or project can properly evaluate itself, and therefore evaluation staff must be independent of project management in technical direction and analysis methods; although of course they should work closely with project staff for administrative purposes.

The project's technical basis and assumptions

The modernisation of a static agricultural system always implies the injection of a set of measures designed to change the production functions. The aim is to increase the productivity of the factor resources, in this case, primarily land and labour. This requires, in turn, the introduction of a technology tailored to the specific requirements of the project area, and sufficiently attractive financially to farmers to enable them to undertake the risks. We illustrate the various steps and assumptions involved in this case study.

The existing cropping patterns and land use are shown in Table 12.3. The most significant element to note in Table 12.3 is that while the gross area remains constant, at 390 000 ha, substantial increases are projected to occur in the areas cropped with improved practices at the cost of the area covered by traditional practices. For example, in the case of sorghum, the main food crop, the percentage of the cropped area under sorghum with traditional practices would decline from 31 per cent at the beginning to 16 per cent, while the sorghum area under improved practices would rise from zero to 8 per cent in the same period.

Continuing with the example of sorghum, we show the projected rate of uptake of the new technology.

Table 12.5 shows the situation before the project and the annual changes during the project's development period with respect to number of sorghum growers, area in ha, and production. Under each sub-heading,

TABLE 12.5 Nigeria: Funtua agricultural development project – Sorghum production projections

Year 6	Without-project	Year 1 project	Year 2	Year 3	Year 4	Year 5
Growers						
Traditional	100 000	96 000	84 500	70 000	60 000	51 000
Intermediate	–	4 000	16 000	31 000	43 000	57 000
Advanced	–	–	500	1 000	2 000	3 000
	100 000	100 000	101 000	102 000	105 000	111 000
Area (ha)						
Traditional	120 000	115 000	101 000	82 000	70 000	61 200
Intermediate	–	5 000	8 000	16 000	21 000	24 000
Advanced	–	–	1 000	2 000	4 000	6 000
	120 000	120 000	110 000	100 000	95 000	91 200
Production (tonnes)						
Traditional	90 000	86 250	75 750	61 500	52 500	45 900
Intermediate	–	7 500	12 000	24 000	31 500	36 000
Advanced	–	–	2 500	5 000	10 000	15 000
	90 000	93 750	90 250	90 500	94 000	96 900
Average yield (kg/ha)	750	781	820	905	989	1 063

growers are grouped in three categories, namely – traditional, intermediate and advanced, the latter representing the most progressive group of farmers who would adhere most closely to the recommended technical package. The more important observations from the Table 12.5 include the following. The total number of sorghum growers would increase by 11 per cent over the period, but the total cropped area would decline substantially by roughly 25 per cent; while yields, expressed in kg/ha, would rise substantially from 750 kg to 1 063 kg in the same period. The net impact of all these changes on total sorghum production would be an additional production of 6 900 tonnes in the same period. While this aggregate change on total production appears modest, the significant element to note is that the productivity of the land resource devoted to it is considerably increased, while the area saved by reducing plantings of sorghum would be converted to maize cultivation.

In contrast, in the case of cotton, for which a more fully tested technology existed, the number of cotton growers would decrease by about 12 per cent, the total cotton-cropped area would increase by about 10 per cent, but average yield would rise from 266 kg to 603 kg/ha. The potential for a more than doubled seed-cotton yield, with modest changes in number of growers and hectarage, makes possible, substantial increases in aggregate seed-cotton production, which was projected to rise from about 24 000 tonnes to about 80 000 tonnes in the same five-year period.

The existing situation for sorghum, the recommended practices, based on research findings at Daudawa research station, and the project proposals are shown below.

Sharing with bulrush millet, the role of the traditional grain staple of the Hausa people, sorghum is grown widely throughout the project area. White to yellowish grained, open-headed Guinea race types, known locally as *farafara*, predominate. The grain is normally ground to a coarse flour and eaten as porridge. The stalks, which are seldom less than 4 m in height, are used for a variety of purposes of which building and fencing are among the most important.

Under the traditional system of cultivation, mean yields are said to approximate 750 kg of threshed grain per ha, but they vary widely in response to intensity of intercropping, soil and climatic effects. No data are available on sorghum production in the project area. Nevertheless, it is safe to conclude that the crop is grown on all holdings, alone or in a variety of mixtures.

Available studies indicated that sorghum occupied approximately 30 per cent of total area planted and, whilst most commonly grown in a two-crop combination with bulrush millet, was also planted alone or with up to four additional crops. Using this as a basis, it was calculated that some 120 000 ha of sorghum are planted in the project area which, at an average yield of 750 kg/ha, would produce some 90 000 tonnes or about 9 per cent of estimated production from North Central State.

Although its effects have not been quantified, the parasitic weed *Striga hermontheca* is probably the most significant pest of sorghum. Others include a range of stalk borers, but particularly *Busseola fusca* and sorghum midge *Contarinia sorghicola*. Major diseases include downy mildew, *Sclerospora sorghi*, a variety of leaf diseases causing spotting and striping, and a number of smut diseases, mainly of the genus

Sphacelotheca. The latter are to a considerable extent controlled by seed dressing with copper fungicides which are readily available commercially.

As befits a staple grain, sorghum is normally planted on the main farm in close proximity to the homestead. Prior to planting, land is ploughed and ridged by hand-hoe or occasionally mechanically, at approximately 0·9 m between rows and varying distances within rows according to intercropping practices. Sorghum is planted in the project area in May/June, after such crops as early millet. Two or three weedings and earthings-up are normally carried out. The crop flowers at the onset of the dry season, in late September/early October, and matures, utilising residual soil moisture, by late November/December.

The stalks are cut near ground level and stacked in the field after removal of the heads, which are bundled and stored in mud-plastered granaries (*rumbus*) in the farm compound. Heads are removed from the granaries as required for domestic consumption or for sale, following threshing. Storage losses are variable, one estimate is that on average 20 per cent of the crop is lost in store, and though control methods involving granary sanitation and the use of insecticides have been developed, they are not commonly used.

Recommended practices

Research on sorghum has been carried out in the northern states of Nigeria for the past 20 years. Higher-yielding varieties have been selected from local types. A number of dwarf varieties and hybrids have also been developed. This work has been accompanied by investigations including time of planting, spacing and fertiliser responses. Improved local varieties grown under improved agronomic practices showed yield gains over traditional types of six to 25 per cent (averaging 18·6 per cent or about 240 kg/ha) in over 90 trials planted at different sites in main sorghum production areas up to 1970. Such increases are of little interest to farmers. The main thrust of sorghum research is now directed towards production of high-yielding dwarf varieties and hybrids. Material now available has a yield potential under field conditions of around 2 500 kg/ha from dwarf varieties and hybrids respectively. The grain from these types is not yet widely acceptable for human food in rural areas but it can be sold readily in urban areas for human and livestock food.

Project proposals

Total sorghum production in Nigeria has been estimated to approximate four million tonnes per year, of which between five and ten per cent is marketed. Prices have risen overall in the early 1970s. However, any major production increase could be expected to result in a marked fall in price – a situation unlikely to arise in the near future in the light of the ineffectiveness of existing extension, input supply and credit services in the northern states as a whole. Within the proposed project area, however, there are good opportunities for promoting some increase in sorghum production, utilising available technology and proposed project services, despite the fact that many farmers are likely to continue to grow at least some local sorghum under traditional cultivation methods. In addition to the switch from traditional to improved varieties of sorghum that was envisaged, it was also believed that a substantial amount of land which in the past had been used to grow sorghum would, under the project, be used for maize.

Two levels of improved technology were assumed, based upon results from field trials and research recommendations: intermediate and advanced (Table 12.6).

TABLE 12.6 Nigeria: Funtua agricultural development project – Projected adoption rates of proposed technology

	Level of technology		
	Traditional	**Intermediate**	**Advanced**
Variety	local	dwarf	hybrid
Plant population/ha	19 200	34 800	48 000
Fertiliser application[1] (kg/ha)	nil	100 kg of calcium ammonium nitrate + 100 kg single superphosphate	200 kg of calcium ammonium nitrate + 100 kg single superphosphate
Borer control[2]	nil	nil	nil
Anticipated yield of threshed grain (kg/ha)	750	1 700	2 500

[1] At the advanced level of technology, two-thirds of the nitrogenous fertiliser would be applied as a side dressing about four weeks after planting. The other fertiliser would be broadcast and worked in before planting.
[2] No recommendations were currently available; insecticidal dusting with lindane or its equivalent may well prove to be economic, however.

In addition to the major elements indicated, farmers would be advised on field preparation, planting methods and other improved practices by project extension staff.

The effects at the farm level of these proposed project actions must be demonstrated somehow. In this case, the absence of basic data on farming systems and the profitability from individual crops, prevents an assessment of the project's effects on farm-family income, and instead of attempting to formulate model farm budgets, which would not be realistic, it was decided instead to express the data in comparative terms, for the various major crops, showing net returns per ha, and net return to labour, based upon the varying levels of technology, namely, traditional, intermediate and advanced. Table 12.7 shows details on net returns per level of technology.

In the case of sorghum, the basic assumptions are elaborated in Table 12.8 for the budget (expressed in Nairas) for one hectare of sorghum.

The budget shows the impact of costs and returns of farmers employing various levels of technology, and expresses this also in incremental terms. The main items of expenditure are shown for labour and for crop inputs. For a farmer, moving from the traditional to the intermediate technology, the additional labour requirements would be 13 man-days, and his incremental net returns per ha would be ₦44·92. The incremental returns per man-day would thus be ₦3·46, which would be a sufficient incentive to induce him to adopt the new technology.

TABLE 12.7 Nigeria: Funtua agricultural development project – Grower returns based on level of technology

	Level of technology		
	Traditional	Intermediate	Advanced
Net return ₦/ha			
With all labour costed			
Cotton	24·52	69·81	110·30
Groundnuts	4·64	25·68	–
Sorghum	31·90	71·62	139·38
Maize	38·25	73·90	182·50
Cowpeas	(9·50)	15·50	–
Net return ₦/ha			
All labour costs excluded			
Cotton	47·72	107·01	157·50
Groundnuts	43·04	75·68	–
Sorghum	65·90	110·82	187·38
Maize	75·85	120·30	234·50
Cowpeas	16·50	48·70	–
Net return to labour (₦/man-day)			
Cotton	0·82	1·15	1·33
Groundnuts	0·45	0·01	–
Sorghum	0·78	1·13	1·56
Maize	0·81	1·04	1·80
Cowpeas	0·25	0·59	–
Net incremental return to incremental labour requirements (₦/man-day)			
Cotton		1·69	1·83
Groundnuts		1·13	–
Sorghum		3·46	3·47
Maize		2·02	4·41
Cowpeas		1·79	–

Project cost estimates and financing

Project costs were estimated to total ₦37·9 million (US$57·6 million) with a foreign exchange component of ₦19·1 million (US$29·2 million) or 51 per cent of total costs. The summary of project costs is shown in Table 12.9.

The following points should be noted by the reader. Costs are presented by major sub-headings that reflect the project's main components, broken down further by items of expenditure for each component. The presentation of costs on an annual basis over five years reflects the proposed phasing or rate of development of the project. One very important index is the percentage distribution of the various components as shares of total costs. Thus, the farm-inputs component accounts for 30 per cent of total project costs because it is the single most important requirement in implementing the new technology. Only incremental requirements were costed in deriving the costs of farm inputs. Participating farmers would receive a 50 per cent subsidy on the cost of phosphatic and nitrogenous fertilisers and the cost of this subsidy is included in project costs, since it involves a financial cost that would be borne by the government.

TABLE 12.8 Nigeria: Funtua agricultural development project – Budget for one hectare of sorghum

	Actual						Incremental			
	I[1]		II		III		II		III	
Yield threshed grain (kg/ha)		750		1 500		2 500		750		1 750
Gross return at ₦0·090/kg		67·50		135·00		225·00		67·50		157·50
Expenditure (₦)										
Labour[2]										
Cultivation	(73)	29·20	(76)	30·40	(84)	33·60	(3)	1·20	(11)	4·40
Fertiliser applications	–	–	(5)	2·00	(12)	4·80	(5)	2·00	(12)	4·80
Harvesting	(12)	4·80	(17)	6·80	(24)	9·60	(5)	2·00	(12)	4·80
Sub-total	(85)	34·00	(98)	39·20	(120)	48·00	(13)	5·20	(35)	14·00
Crop inputs[3]										
Tools		1·00		1·00		1·00		–		–
Seed[4]		0·60		1·58		4·22		0·98		3·62
Fertiliser: N[5]		–		10·80		21·60		10·80		21·60
P_2O_5[6]		–		10·80		10·80		10·80		10·80
Sub-total		1·60		24·18		37·62		22·58		36·02
Total expenditure		35·60		63·38		85·62		27·78		50·02
Net returns										
With labour		31·90		71·62		139·38		39·72		107·48
Without labour		65·90		110·82		187·38		44·92		121·48
Per man-day		0·78		1·13		1·56		3·46		3·47

1 Farmer category: I = traditional with no improved inputs; II = intermediate using fertiliser and improved dwarf sorghum from local varieties; III = advanced farmers using higher application of nitrogen and in some instances hybrid seed.
2 Labour costed at shadow wage rate for family and hired labour at ₦0·4 per man-day (6 h) – ( ) indicates number of man-days.
3 Includes interest rates at 10%, except for tools in category II and category III.
4 Category II seed, 12 kg/ha at ₦0·132/kg, category III seed, 16 kg/ha at ₦0·264/kg.
5 Calcium ammonium nitrate category II 100 kg/ha, category III 200 kg/ha subsidised cost ₦108·0/tonne.
6 Single superphosphate category II and III 100 kg/ha subsidised cost ₦100·8/tonne.

TABLE 12.9 Nigeria: Funtua agricultural development project – Summary of project costs (₦'000)

	Year 1	Year 2	Year 3	Year 4	Year 5	Total	Foreign exchange per cent	Foreign exchange cost	Local cost	Local per cent[1] distribution
Project administration										
Vehicles and equipment	43·1	33·5	1·6	14·0	1·6	93·8	61	56·8	37·0	
Houses, buildings and furnishings	246·0	282·0				528·0	30	160·4	367·6	
Salaries and allowances	118·9	166·4	176·9	176·9	161·9	801·0	28	223·0	578·0	
Vehicles operating	4·5	6·3	6·3	6·3	6·3	29·7	44	13·0	16·7	
Head Office expenses (GS)	62·6	92·8	139·6	173·1	176·5	644·6	21	136·1	508·5	
Sub-total	475·1	581·0	324·4	370·3	346·3	2 097·1	28	589·3	1 507·8	6
Technical services										
Vehicles and equipment	203·8	86·1	65·3	59·4	44·2	458·8	71	324·4	134·4	
Houses, buildings and furnishings[2]	614·0	640·0	801·0	57·0		2 112·0	8	587·3	1 524·7	
Salaries and allowances	183·6	322·3	440·2	483·9	393·9	1 823·9	22	398·0	1 425·9	
Vehicles operating	24·8	42·1	45·8	48·1	48·1	208·9	44	91·9	117·0	
Tractor hire service (GS)	20·3	16·2	12·2	8·1		56·8	44	25·0	31·8	
Farm training centres operating (GS)	8·0	16·0	16·0	16·0	16·0	72·0	–	–	72·0	
Seed-multiplication farms (GS)	73·0	73·0	73·0	73·0	73·0	365·0	50	182·5	182·5	
Sub-total	1 127·5	1 195·7	1 453·5	745·5	575·2	5 097·4	31	1 609·1	3 488·3	13
Farm-credit services										
Vehicles and equipment	117·5	73·0	157·6	66·5	55·6	470·2	65	306·0	164·2	
Houses and furnishings	274·0	179·0	270·0			723·0	20	145·8	577·2	
Salaries and allowances	121·7	184·7	299·4	299·4	299·4	1 204·6	10	115·0	1 089·6	
Vehicles operating	15·3	19·3	30·7	30·7	30·7	126·7	44	55·7	71·0	
Credit operations (GS)	0·5	2·0	3·5	5·0	8·0	19·0	60	11·4	7·6	
Sub-total	529·0	458·0	761·2	401·6	393·7	2 543·5	25	633·9	1 909·6	7

Market services										
Vehicles and equipment	52·5	42·0	86·0	22·5	2·0	205·0	76	154·8	50·2	
Houses, furnishings and ginnery market depots	128·0	203·0	94·0			425·0	29	123·0	302·0	
Salaries and allowances	88·5	121·0	156·1	156·1	133·1	654·8	14	92·0	562·8	
Vehicles operating	12·1	13·1	13·1	13·1	13·1	64·5	44	28·4	36·1	
Sub-total	281·1	379·1	349·2	191·7	148·2	1 349·3	30	398·2	951·1	4
Engineering services										
Overheads										
Vehicles and workshop equipment	109·0	13·9	13·9	30·5	0·7	168·0	73	123·4	44·6	
Houses and furnishings	349·0	36·0				385·0	30	115·5	269·5	
Salaries and allowances	178·3	188·3	188·3	169·5	101·3	825·7	45	370·0	455·7	
Vehicles operating	13·5	18·3	18·3	18·3	18·3	86·7	44	38·2	48·5	
Workshop operating	24·8	44·7	44·7	44·7	44·7	203·6	49	99·4	104·2	
Sub-total	674·6	301·2	265·2	263·0	165·0	1 669·0	45	746·5	922·5	4
Road construction										
Plant and equipment	236·9	134·3	16·4	1·0	1·0	389·6	89	346·8	42·8	
Operating plant and equipment	83·1	153·2	153·2	153·2	153·2	695·9	38	264·0	431·9	
Construction materials, labour	76·0	73·0	73·0	73·0	108·0	403·0	54	218·9	184·2	
Land compensation (GS)	18·0	18·0	18·0	18·0	18·0	90·0			90·0	
Road maintenance (GS)	121·0	38·3	38·3	50·3	38·3	286·2	53	151·0	135·1	
Sub-total	535·0	416·8	298·9	295·5	318·5	1 864·7	53	980·7	884·0	5
Soil and water conservation										
Plant and equipment	278·2			29·4		307·6	86	265·2	42·4	
Operating plant and equipment	14·8	104·8	104·8	104·8	104·8	434·0	42	180·7	253·3	
Construction materials, labour	34·0	90·0	90·0	90·0	90·0	394·0	48	188·1	205·9	
Land compensation (GS)	2·0	8·0	8·0	8·0	8·0	34·0			34·0	
Sub-total	329·0	202·8	202·8	232·2	202·8	1 169·6	54	634·0	535·6	3
Total engineering services	**1 538**·6	920·8	796·7	760·9	686·3	4 703·3	50	2 361·2	2 342·1	(12)

TABLE 12.9 (continued)

	Year 1	Year 2	Year 3	Year 4	Year 5	Total	Foreign exchange per cent	Foreign exchange cost	Local cost	Local per cent[1] distribution
Farm inputs										
Fertiliser subsidy (gross)	170·5	424·4	978·6	1 600·3	2 091·9	5 265·7	72	3 791·3	1 474·4	14
Incremental fertiliser inputs	170·4	255·1	562·2	641·0	506·3	2 135·0	72	1 537·2	597·8	6
Incremental seeds, insecticides and batteries	127·9	289·6	504·5	461·9	294·1	1 678·0	60	1 006·8	671·2	4
Incremental farm equipment	92·6	266·5	558·4	685·2	746·2	2 348·9	79	1 855·6	493·3	6
Sub-total	561·4	1 235·6	2 603·7	3 388·4	3 638·5	11 427·6	72	8 190·9	3 236·7	30
Evaluation unit[3]	483·1	245·8	250·6	272·8	250·6	1 502·9	49	731·9	771·0	4
Total	4 995·8	5 016·0	6 509·5	6 161·0	6 028·8	28 721·1	51	14 514·5	14 206·6	76
Physical contingencies[4]	249·8	250·8	325·5	308·1	301·9	1 436·1	51	732·4	703·7	4
Price contingencies	901·0	1 322·0	1 980·0	1 670·0	1 836·0	7 709·0	51	3 391·6	3 777·4	20
Total project costs	6 146·6	6 588·8	8 815·0	8 139·1	8 176·7	37 866·2	51	19 178·5	18 687·1	100
Percentage distribution	16	17	23	22	22	100		51	49	
US$	9 342·8	10 015·0	13 398·8	12 371·4	12 428·6	57 556·6		29 151·3	28 405·0	

(GS) indicates general service costs

[1] Total distribution – administrative overheads 6%, direct farmer services 25%, roads, water and soil conservation 14%, farm inputs 22%, evaluation 4%.

[2] Includes village service centres, farm training centres and seed multiplication.

[3] Includes evaluation section of PMEU at Kaduna to service Funtua, Gombe and Gusau and field enumeration unit for Funtua.

[4] Physical contingencies at 5% per annum.

Extension, credit and marketing services are second in the cost ranking with 24 per cent of total project costs with roads, soil and water conservation occupying third place with 12 per cent of total costs. Project administration would account for 6 per cent – which is a reasonable share for a project of this magnitude – while the evaluation unit would account for 4 per cent. The above costs all total 76 per cent of total project costs. The remaining 24 per cent of total project costs includes 4 per cent for physical contingencies and 20 per cent for price contingencies, over the proposed five-year disbursement period. The price contingencies did not include provision for escalation of fertiliser prices, since fertiliser costs had already risen sharply by 1974 in comparison with others and average current 1980 prices for fertilisers were projected to be about the same as current 1974 prices.

The method of derivation of costs for the water and soil conservation component is shown in Table 12.10.

It is seen that costs over the five-year period total ₦1·135 million. To this must be added ₦0·034 million – the costs of land acquisition – to derive the total costs shown in the summary table of project costs for this component. The use of unit quantities and unit costs should be noted in the derivation of these cost estimates.

The financing package agreed upon for this project was as follows: a World Bank loan of US$29 million to the Federal Military Government of Nigeria, which equalled 51 per cent of total project costs and covered all the foreign exchange requirements of the project. Of this amount, the FMG would on-lend ₦18·4 million (US$27·9 million) to the North Central State government, while the FMG would retain US$1·1 million to finance the evaluation section of the project monitoring and evaluation unit, and would, on its own account, contribute ₦0·4 million (US$0·7 million) to the financing of the evaluation unit. The government of North Central State would contribute ₦16·4 million (US$25 million) that is 43 per cent of total project costs to the project's investment and operating costs. The state government would bear all the costs of the fertiliser subsidy estimated at ₦5·26 million. The contribution of participating farmers would amount to ₦1·9 million (US$2·9 million) or about 5 per cent of project costs in the form of cash purchases of seasonal inputs and down payments on farm equipment.

Organisation and management (O and M)

There are two organisational levels involved in management of the project: the Federal Military Government and the Project Management Unit.

The Federal Military Government is involved in the project through its overall responsibility for channelling the external loan resources to the central state government; also a small financial contribution is made to the operations of the project monitoring and evaluation unit. More significant, however, is the fact that the various state-level ministries, departments and agencies must be guided in varying degrees, by the overall policy directives and programmes that are being promoted at the national level. For example, there are such national policies relating to subsidies on inputs, marketing arrangements and pricing formulas for

TABLE 12.10 Nigeria: Funtua agricultural development project – Water and soil conservation development costs (Naira)

	Unit	Unit cost	Number	Year 1	Year 2	Year 3	Year 4	Year 5	Total	Foreign exchange per cent	Foreign exchange cost	Local cost
Plant, vehicles and equipment												
Tractor dozer/ripper 140 hp	No.	32 800	1	(1) 32 800			–		32 800	90	29 520	3 280
Tractor dozer/ripper 180 hp	No.	51 900	1	(1) 51 900			–		51 900	90	46 710	5 190
Self-propelled scraper 150 hp	No.	37 600	2	(2) 75 200			–		75 200	90	67 680	7 520
Motor grader 125 hp	No.	29 400	1	(1) 29 400			–		29 400	90	26 460	2 940
Agricultural tractor 65 hp	No.	4 000	4	(4) 16 000			–		16 000	90	14 400	1 600
Compactor	No.	4 000	1	(1) 4 000			–		4 000	90	3 600	400
Water bowser	No.	1 000	3	(3) 3 000			–		3 000	90	2 700	300
Fuel bowser	No.	4 500	2	(2) 9 000			–		9 000	90	8 100	900
Concrete mixer	No.	1 000	2	(1) 1 000			(1) 1 000		2 000	90	1 800	200
Water pumps	No.	500	8	(4) 2 000			(4) 2 000		4 000	90	3 600	400
Generator 2·5 Kw	No.	1 000	1	(1) 2 500			–		2 500	90	2 250	250
Caravan	No.	3 000	5	(5) 15 000			–		15 000	90	13 500	1 500
Crew vehicle 4 wd	No.	4 000	6	(3) 12 000			(3) 12 000		24 000	90	21 600	2 400
Pick-up trucks ½ tonne	No.	3 200	4	(2) 6 400			(2) 6 400		12 800	60	7 680	5 120
Lorry 7 tonne	No.	8 000	2	(1) 8 000			(1) 8 000		16 000	60	9 600	6 400
Survey equipment	Set	10 000	1	(1) 10 000			–		10 000	60	6 000	4 000
Sub-total				278 200			29 400		307 600	86	265 200	42 400
Construction materials and labour												
Concrete	Per dam	1 000	85	(5) 5 000	(20) 20 000	(20) 20 000	(20) 20 000	(20) 20 000	85 000	–	–	85 000
Fencing	Per dam	1 000	85	(5) 5 000	(20) 20 000	(20) 20 000	(20) 20 000	(20) 20 000	85 000	50	42 500	42 500
Pipes, fittings, etc.	Per dam	2 000	85	(5) 10 000	(20) 40 000	(20) 40 000	(20) 40 000	(20) 40 000	170 000	80	136 000	34 000
Tents and camping gear	Set	12 000	1	(1) 12 000	–	–	–	–	12 000	80	9 600	2 400
Labour	Man-days		42 000	2 000	10 000	10 000	10 000	10 000	42 000	–	–	42 000
Sub-total				34 000	90 000	90 000	90 000	90 000	394 000	48	188 100	205 900
Plant and equipment operating												
Tractor dozer/ripper 140 hp	Hours	6·19	8 240	1 486	12 380	12 380	12 380	12 380	51 006	34	17 342	33 664
Tractor dozer/ripper 180 hp	Hours	9·01	8 240	2 162	18 020	18 020	18 020	18 020	74 242	37	27 470	46 772
Self-propelled scraper 150 hp	Hours	7·89	16 480	3 787	31 560	31 560	31 560	31 560	130 027	44	57 212	72 815
Motorgrader 125 hp	Hours	7·19	8 240	1 726	14 380	14 380	14 380	14 380	59 246	44	26 068	33 178
Agricultural tractor and compactor 65 hp	Hours	2·51	24 720	602	5 020	5 020	5 020	5 020	20 682	44	9 100	11 582
Agricultural tractor (with ploughs)	Hours	1·39	16 800	1 001	8 340	8 340	8 340	8 340	34 361	44	15 119	19 242
Crew vehicle 4 wd	Hours	1·62	16 800	1 944	6 318	6 318	6 318	6 318	27 216	44	11 975	15 241
Pick-up truck ½ tonne	Hours	1·40	11 000	840	3 640	3 640	3 640	3 640	15 400	44	6 776	8 624
Lorry 7 tonne	Hours	2·34	8 300	702	4 680	4 680	4 680	4 680	19 422	44	8 546	10 876
Other	Per annum	500		500	500	500	500	500	2 500	44	1 100	1 400
Sub-total				14 750	104 838	104 838	104 838	104 838	434 102	42	180 708	253 394
Total				326 950	194 838	194 838	224 238	194 838	1 135 702		634 008	501 694

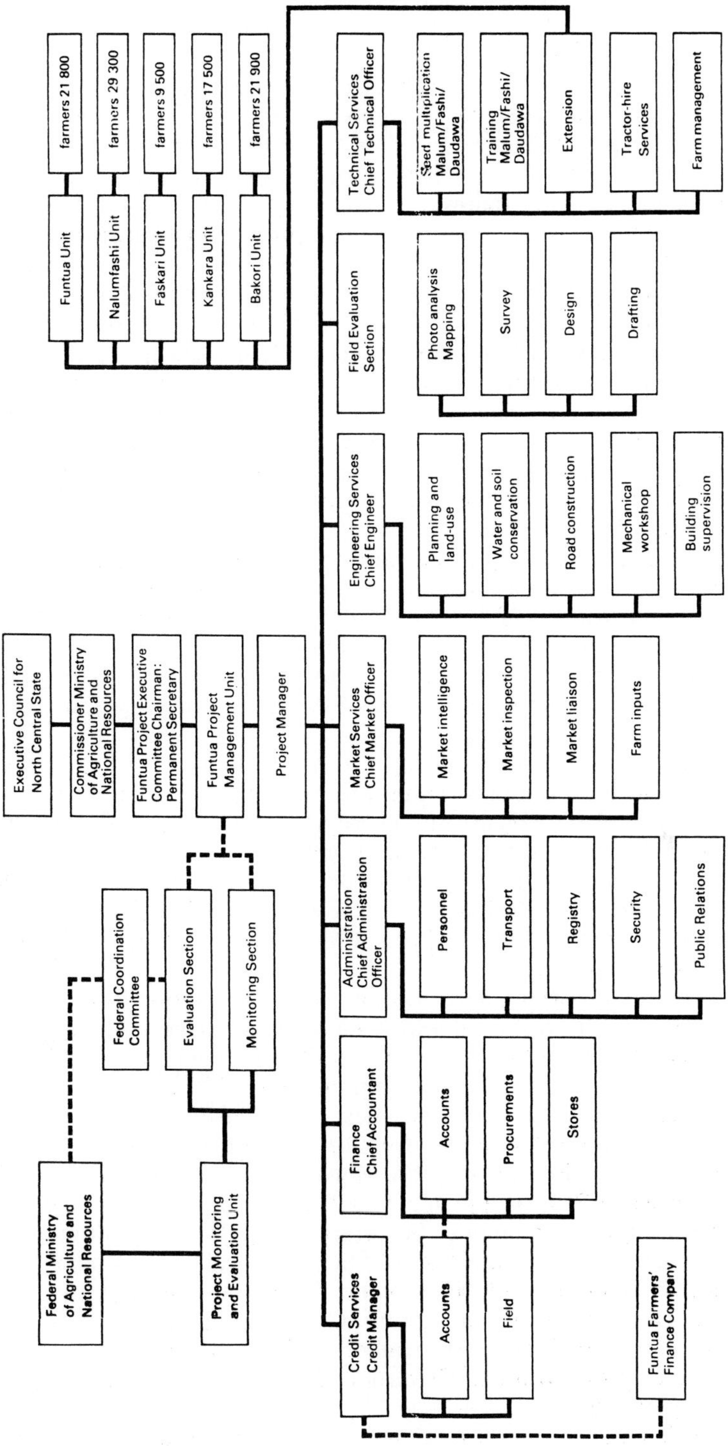

Figure 12.1 Funtua agricultural development project – Project organisation

agricultural produce that impinge heavily on the project's performance. The various federal agencies also have their own programmes for various states.

The project's organisation structure must, therefore, reflect the federal state government relationships, as shown in the project's organisation chart (Figure 12.1). In order to ensure coordination, it is noted that a Federal Coordinating Committee was established, whose composition appears below. The presence of a Federal Ministry of Finance representative is noted, because project financing arrangements are always critical. The broken lines of the chart show that the project monitoring and evaluation unit, while federally administered, stands in a staff relationship *vis-à-vis* the project's main implementing authority.

Primary responsibility for project implementation would be vested in a specially constituted Funtua Project Management Unit (FPMU), which would exercise line authority in the project's management. Apart from the coordination arrangements at the federal level, proper organisational arrangements must be made to ensure coordination at the state level. In this case, the project provided for this by the creation of the Funtua Project Executive Committee (FPEC), whose composition reflects the involvement of the various North Central State agencies in the project's execution. This stems, in turn, from the project's main components, for example, the Ministry of Transport and Works is involved because of the road component. This new entity would have to be endorsed with a certain degree of autonomy, as one means to minimise the impact of the bureaucratic processes which tend to slow down the rate of project implementation.

The other organisational layer is that dealing with the project itself and internal arrangements within the project's management unit. The organisation chart (Figure 12.1) depicts this clearly. For example, provision is made for engineering, marketing, credit and technical services geared to supporting the farmers, e.g. training and extension. In addition, there are the usual finance and administration aspects, typical of all projects. One person, the Project Manager, has responsibility for all these various functions.

The organisation chart must, however, be supported by a written description of the responsibilities. Thus, the managerial and technical responsibilities of the head and senior staff of each unit must be spelt out.

Funtua Project Executive Committee would, *inter alia*, determine policy, coordinate project activities with those of associated state ministries and departments, exercise budgetary and audit control, approve the appointment of professional grade staff and review and set salary scales. FPEC would meet at least once a quarter and would comprise: the Permanent Secretary MANR as chairman, and representatives of the Ministries of Finance, Local Government, Works, Economic Planning and NCS Water Board, a representative of the Federal Department of Agriculture, Chief Agricultural Officer for the North Central State, and two members residing in the project area to be appointed by the Military Governor. The Project Manager would be secretary of FPEC.

A Federal Coordinating Committee would be established to review the progress of the Funtua and similar projects in the northern states. It would provide coordination between other Bank-financed agricultural projects of a similar nature in northern Nigeria, and would ensure a full exchange of ideas and results. In its review of project progress, it would be aided by the evaluation section of the project monitoring and evaluation unit (PMEU). The committee would meet at least twice a year and would comprise: the Director of the Federal Department of Agriculture (FDA) as chairman; Chief Agricultural Planning Officer (FDA) as secretary; a representative from the respective project executive committees; Project Managers of Funtua, Gombe and Gusau; Director of the Institute of Agricultural Research, the Head of the Evaluation Section of PMEU, representatives of NAB and NPMC; and representatives of the Federal Ministry of Finance.

Project administration would fall under the direct responsibility of the Project Manager for all operational procedures and actions. He would be responsible to FPEC. The FPEC, in consultation with the Project Manager, would appoint one of the senior project staff as Deputy Project Manager. The Project Manager would be assisted by: Chief Administrative Officer, Chief Accountant, Chief Technical Officer, Chief Engineer, Chief Marketing Officer, and Credit Manager. A summary showing the functions of the project's six operational sections follows.

Administration Section would be responsible for all projects personnel matters, staff transportation, registry, security and public relations.

Finance Section would be responsible for project accounts, procurement, stores, internal audit, budget control and would set up systems to provide accurate cost accounts for evaluation and management purposes.

Technical Services Section would be responsible for farm extension advice, staff and farmer training, seed multiplication, farm-management advice to larger farmers and tractor-hire services. The main development effort would be through the extension service and structurally it would consist of five development unit centres based on the district towns of Funtua, Malumfashi, Faskari, Kankara and Bakore, each headed by a development officer, who would supervise agricultural assistants (AA) posted to farm service centres in each of the 77 villages of the project area. Each AA would supervise an average of four agricultural instructors (AI) who would live in the hamlets associated with each village and who would work with the hamlet farmers and their committees. In this way, the extension system would be tailored to the traditional village structure.

Credit Services Section would work parallel with the extension services. Credit officers would be posted to Development Unit Centres, and two credit assistants to each farm service centre.

Market Services Section would monitor the existing marketing system, liaise and provide market information for buyers and farmers, control the activities of licensed buying agents, and supervise the work of produce examiners. A marketing supervisor would be assisted by two marketing assistants posted to each Development Unit Centre and a produce examiner would be posted to each market. Market services would be responsible for the logistics involved in procurement of farm inputs and for their distribution and handling. As such, the produce examiner would carry out the functions of store keeper when the markets are closed.

Engineering Services Section would be a specialised service responsible for land-use planning, road construction, water development, soil conservation, building supervision and workshop management. It would prepare tender documentation for procurement of engineering equipment and supplies, supervise building contractors, and implement all construction aspects of the project not undertaken by contractors. It would maintain project vehicles and equipment.

Field evaluation would be directed by the evaluation section of the project monitoring and evaluation unit (PMEU). The field enumeration staff would be administratively responsible to the Project Manager.

The constitution of these bodies, namely the Funtua Project Management Unit (FPMU) and the Funtua Project Executive Committee (FPEC) must have a legal basis. In this case, the publication by the North Central State Executive Council of an 'order' in the State Government Gazette establishing these two organs constituted an adequate legal basis. The objectives, composition and authority relationships of these bodies as proposed at appraisal appear below.

The FPMU would be managed by a Project Manager and its objectives would be as follows.

1 Development of Funtua Adminstrative Division with a view to increasing production of agriculture and raising the living standards of the rural population.
2 Construction of roads, water supplies and soil-conservation works associated with agricultural development.
3 Development and management of seed-multiplication farms.
4 Training of project staff and farmers.
5 Provision of village service centres and stores.
6 Provision of extension, credit and marketing services.
7 Procurement of farm inputs.
8 Development of supporting infrastructure such as houses, buildings, offices, etc., for the above services.

The FPMU would be controlled by the Funtua Project Executive Committee (FPEC) which would be made up as follows.

Permanent Secretary – Ministry of Agriculture and Natural Resources
Representative – Ministry of Finance
Representative – Ministry of Local Government
Representative – Ministry of Transport and Works
Representative – Ministry of Economic Planning
Representative – NCS Water Board
Representative – North Central State Works Board
Director of Federal Agricultural Department or representative
Chief Agricultural Officer, for the North Central State
Two members from the project area to be appointed by the Military Governor.

The Project Manager would be the secretary of the FPEC.

FPEC would be responsible to the Commissioner of Agriculture and would have the following responsibilities:

Policy
FPEC would determine agricultural and development policy, and would consult with the Federal Government and the World Bank when any major changes in policy may be required.

Finance
1 FPEC would approve the annual budget and would ensure that sufficient government funds are readily available to the project.
2 FPEC would appoint independent auditors.
3 FPEC would submit an audited statement of accounts to the Commissioner of Agriculture within four months of the closure of each financial year.
4 FPEC would approve the opening of project bank accounts.
5 The chairman of FPEC would approve the monthly cash-flow statements submitted by the manager of FPMU.
6 The chairman would approve annually the terms of lending to project farmers on the basis of recommendations made by FPMU.

Staff and personnel matters In consultation with the manager of FPMU, EPEC would be responsible for the appointment of all senior project staff and employees, their terms and conditions of employment and annual review of salaries and allowances.

Award of contracts FPEC would on the advice of the manager of FPMU be responsible for approval of any single contract valued at more than ₦30 000.

Inter-departmental coordination The FPEC through its members, would be responsible for the coordination at ministry level of those issues affecting the proper development of the project.

Meetings of FPEC

1 The chairman would preside over all the meetings of the FPEC which he attends, and in his absence the present members would appoint one of their number to preside over such a meeting.
2 The chairman of the FPEC would invite the FPEC to meet at a suitable time and place; provided that no more than three months would elapse between any meeting and the next. In the absence of the chairman, the manager would call the FPEC to meet whenever he deems this necessary.
3 Six members of the FPEC would constitute a quorum of any meeting of the FPEC. The FPEC would pass its decisions by the majority of the votes of its members present and voting, and in case of equality of votes the chairman would have a casting vote.
4 The FPEC may call any person to attend any of its meetings without that person having a counted vote.

FPMU would be responsible to FPEC for the day-to-day operations of the project. In particular it would be responsible for the carrying out of the objectives as stated above. FPMU would be managed by a Project Manager who would report directly to the chairman of FPEC. The main duties of FPMU would be as follows.

1 The implementation of an agricultural development programme within Funtua Administration Division following the policy guidelines of the FPEC.
2 The control and operation of the project's revolving-credit fund.
3 The preparation of annual budgets and cash flows for approval by FPEC. In particular FPMU would submit to the chairman of the FPEC monthly cash-flow statements indicating any changes in requirements for state government funds due to changes in project expenditures and inflow anticipated from the federal government. Unless informed to the contrary within five days of dispatch of such statements to the chairman of the FPEC, the manager would be authorised to make the proposed expenditures.
4 The recruitment of project staff, in consultation, when necessary, with other government departments, and senior staff, with approval of FPEC.
5 The advertising, negotiating, calling for tenders and after approval from FPEC, for concluding such contracts with persons and organisations either inside or outside the State for the erection of buildings, the construction of civil works and the supply of goods, services and materials as may seem to be requisite for the proper discharge of its duties.
6 The operation of the project's bank account.
7 The preparation of reimbursement claims against the project's expenditure, for submission to the federal government and the World Bank.

FPMU would administer a field evaluation unit that would be technically directed by a Federal government-controlled project monitoring and evaluation unit (PMEU).

FPMU would prepare quarterly and annual reports for the simultaneous submission to FPEC, North Central State Executive Council, Federal Government of Nigeria and the World Bank.

FPMU would be delegated any additional duties and powers by FPEC to ensure the attainment of the objectives.

Project's output, prices and financial analysis

The project's contribution to incremental production stemming from project investments is primarily a function of yield assumptions about productivity increases from the technical packages proposed and the rate of adoption, that is, the total hectarage and number of farmers brought in under the project annually over the project's development period. These were discussed in the section dealing with the technical basis. Assumptions about these two parameters are summarised in Table 12.11 on a crop basis.

However, a major determinant of the rate of adoption of the project's proposed package is the incentive aspects of the marketing process and the pricing levels. With regard to the marketing of food crops, the project would introduce improvements in the presently defective system through provision of facilities for assembling and short-term storage at the service centres to farmers' associations. Futhermore, the project would develop a market intelligence system that would keep farmers informed of the supply, demand and prices in various markets.

TABLE 12.11 Nigeria: Funtua agricultural development project: Estimates of production before and after development

	Before development			After development		
	Area (ha)	**Average yield** (kg/ha)	**Total production** ('000 tonnes)	**Area** (ha)	**Average yield** (kg/ha)	**Total production** ('000 tonnes)
Seed cotton	90 540	270	24	99 050	600	60
Sorghum	120 000	750	90	91 250	1 060	97
Maize	5 000	900	5	32 500	2 000	66
Groundnuts	32 500	480	16	35 250	790	28
Cowpeas	3 800	200	1	5 000	670	3

With respect to the cash crops, cotton and groundnuts, positive changes had been introduced in 1973 in statutory marketing of these crops. These included for example, federal responsibility for producer-price determination and support price policies, abolition of producer taxes and the bringing of producer prices more in line with world market prices. In fact, the 1974 producer prices reflected the world market situation and this provided considerable incentives to farmers.

For each of the major crops, it was necessary to project 1980 financial and economic prices. For example, the financial and economic prices are shown in Table 12.12 for maize and sorghum. The derivation of these prices is shown in Table 12.13.

TABLE 12.12 Price projections for 1980 in ₦ per tonne

	Maize	Sorghum
Economic	76·00	74·00
Financial	86·00	90·00

The estimation of these prices took account of the following factors: projected cost of imports; relative production costs; the potential demand for maize for starch and feed industries; the role of sorghum in the diet; and the relative inelasticity of market demand for food grains. It was necessary also to take into account the radius for profitable marketing of the crops having regard to transport costs, that is, the low value/volume ratio of maize and sorghum as compared to cotton and groundnuts.

The project's benefits at the farm level were difficult to quantify for the reasons discussed earlier, namely the unreliability of projecting typical farm budgets. However, the project's actions would generate additional net incomes to all farm families of ₦9·7 million per annum by 1980, that is ₦143 (US$217 per family). This represents an increase of 67 per cent over the pre-project situation. The average *per capita* income would increase from US$41 to US$63 at full development.

At the level of the North Central State government (NCS), the average cost, inclusive of loan servicing, would be about ₦5·5 million annually until 1995. Costs would later be reduced to ₦3·4 million annually. Savings of ₦2·1 million annually could be effected if the government removed the fertiliser subsidy. No tangible financial benefits would accrue to the NCS, since produce taxes on statutory crops, including cotton and groundnuts, had recently been abolished.

There would, however, be benefits to the federal government through secondary or tertiary multiplier benefits, which were estimated to range between 8 per cent and 20 per cent, resulting in indirect taxation revenues of about ₦1·5 million annually.

Economic benefits and justification

The project's main direct benefit would be the incremental production generated, which was estimated to reach N 16·6 million annually at full development by year five onwards. About 60 000 farm families would

TABLE 12.13 Nigeria: Funtua agricultural development project – Maize and sorghum: estimated producer prices (₦ per tonne)[1]

	Maize		Sorghum	
	Financial	Economic	Financial	Economic
f.o.b. US Gulf[2]	59·20	70·90	52·60	63·00
Freight insurance[3]	9·90	9·90	9·90	9·90
c.i.f.	69·10	80·80	62·50	72·90
Duty (40% of c.i.f.)	27·60	–	25·00	–
Handling: Port charges[4]	10·00	10·00	10·00	10·00
	106·70	90·80	97·50	82·90
Inland transport[5]	20·00	15·00	20·00	15·00
Wholesale market value	126·70	105·80	117·50	97·90
Transport from service centre[6]	4·00	3·00	4·00	3·00
	122·70	102·80	113·50	94·90
Local transport, storage, cost of bags	4·00	3·00	4·00	3·00
	118·70	99·80	109·50	91·90
Margins	10·00	10·00	10·00	10·00
	108·70	89·80	99·50	81·90
Storage losses[7]	16·30	13·50	10·00	8·20
'Ceiling' value	92·40	76·30	89·50	73·70
Production cost and 'preference' adjustment	6·00			
Estimated farm-gate price (rounded)	86·00	76·00	90·00	74·00

[1] Financial price bases ₦1 = $1.52. Economic price basis ₦1 = $1.27.
[2] Basis IBRD projected prices in 1974 constant terms, No. 2 US corn $90.00, sorghum $80.00 per tonne.
[3] Basis charter tonnage (12–15 000 tonnes) at US$15.00 per tonne.
[4] Including bagging.
[5] Average 644 km at 3 kobo per tonne per km.
[6] Average 129 km at 3 kobo per tonne per km.
[7] 15% for maize, 10% for sorghum.

TABLE 12.14 Nigeria: Funtua agricultural development project – Project benefits

	Year 1	Year 2	Year 3	Year 4	Year 5	Years 6–25
Incremental production (tonnes)						
Cotton	1 107	7 743	20 429	31 766	35 653	35 653
Sorghum	3 750	(250)	(500)	4 000	6 900	6 900
Maize	–	3 162	12 875	36 125	61 250	61 250
Groundnuts	720	1 680	5 760	8 040	12 120	12 120
Cowpeas	1 115	1 475	1 815	2 165	2 580	2 580
Economic value of incremental production (₦'000)						
Cotton ₦268/tonne	297	2 075	5 475	8 513	9 555	9 555
Sorghum ₦74/tonne	278	(18)	(37)	296	511	511
Maize ₦76/tonne	–	240	979	2 746	4 655	4 655
Groundnuts ₦129/tonne	93	217	743	1 037	1 563	1 563
Cowpeas ₦99/tonne	110	146	180	214	255	255
Total	778	2 660	7 340	12 806	16 539	16 539

TABLE 12.15 Nigeria: Funtua agricultural development project – Economic costs and returns (₦'000)

	Year 1	Year 2	Year 3	Year 4	Year 5	Year 6	Year 7	Year 8	Year 9	Year 10	Years 11–25
Incremental costs											
Farm inputs:											
Fertiliser	341	850	1 966	3 229	4 227	4 227	4 227	4 227	4 227	4 227	4 227
Other seasonal	128	418	922	1 384	1 678	1 678	1 678	1 678	1 678	1 678	1 678
Equipment	93	267	614	869	1 109	201	201	201	201	201	201
Vehicles, plant and equipment[1]	763	383	341	194	105	39	141	67	96	96	96
Housing, building, labour and materials[2]	1 687	1 413	1 238	130	108	(378)	(159)	(31)	(23)	–	–
Salaries	691	983	1 261	1 286	1 090	957	844	769	720	720	720
Vehicle running operations[3]	178	297	312	312	312	188	80	80	80	80	80
General services	305	264	309	352	338	267	254	251	250	250	250
Contingencies at 5%	209	244	347	388	448	377	371	363	362	362	362
Total	4 395	5 119	7 310	8 144	9 415	7 556	7 637	7 605	7 591	7 614	7 614
Less taxes[4]	304	263	254	155	123	91	83	74	72	72	72
Costs adjusted for taxes	4 091	4 856	7 056	7 989	9 292	7 465	7 554	7 531	7 519	7 542	7 542
Add net labour cost	159	103	631	1 306	1 725	1 453	1 439	1 411	1 369	1 313	1 313
Total	4 250	4 959	7 687	9 295	11 017	8 918	8 993	8 942	8 888	8 855	8 855
Add foreign exchange adjustment	390	486	798	1 024	1 245	1 014	1 020	1 008	1 012	1 012	1 012
Net economic costs	4 640	5 445	8 485	10 319	12 262	9 932	10 013	9 950	9 900	9 867	9 867
Incremental benefits											
Farm produce	778	2 660	7 340	12 806	16 539	16 539	16 539	16 539	16 539	16 539	16 539
Seed-farm production	–	124	124	124	124	124	124	124	124	124	124
Net economic benefits	778	2 784	7 464	12 930	16 663	16 663	16 663	16 663	16 663	16 663	16 663
Surplus or deficit	(3 862)	(2 661)	(1 021)	2 611	4 401	6 731	6 650	6 713	6 763	6 796	6 796

1 Does not include replacement for road maintenance equipment (see ground services) excludes water and soil conservation costs.
2 Excludes water and soil conservation costs, includes residual value of vacated staff houses from years 6–9.
3 Excludes water and soil conservation costs.
4 Excludes farm inputs and other relevant contingencies. Identifiable taxes on remainder approximately equals in Year 1, 8·0%; Year 2, 7·5%; Year 3, 7·0%; Year 4, 6·5%, and thereafter 6·0%.

comprise the direct beneficiaries, while a further 40 000 families would enjoy some benefits from the project's infrastructure and marketing arrangements within the project area. When benefits to the non-farming population are included, e.g. traders and labourers, the total beneficiaries of the project's primary and secondary benefits would total about 900 000 people.

The project's internal economic rate of return from investment was estimated to be 41 per cent. Table 12.14 shows the project benefits in terms of incremental production, converted at economic prices to give the economic value of incremental production.

The derivation of the economic rate of return is shown in Table 12.15 which shows the incremental costs adjusted for taxes, the net labour cost for the project (derived by computing the value of incremental labour less labour savings from use of ox-drawn and tractor equipment), and the adjustment for the foreign exchange, which has been priced at ₦1 = US$1·27 against actual rate of ₦1 = US$1·52.

It should be noted that if no adjustment were made by the shadow pricing of foreign exchange, the economic rate of return would drop to 32 per cent. The same table shows the incremental benefits, expressed as net economic benefits, which when matched against net economic costs gives the annual surplus or deficit used to compute the economic rate of return.

The annual costs and returns have taken account of the following assumptions and methods:

1 At full development the area cropped in the five crops tabulated is assumed to increase from 250 000 ha to 263 000 ha. No value has been attributed to the additional land brought under cultivation and no costs included for land acquisition, since this land would be fallow land allocated by traditional rights of occupancy and use.
2 Project life is assumed as 25 years. Most project-development activities will be completed by project-year five (1979), so salvage value is added to benefits in project-years six through nine to account for those houses and buildings no longer required by the project and transferred to other government agencies.
3 A shadow wage rate of ₦0·40 per 6-hour man-day as against the official rate of ₦0·60.
4 The exclusion of identifiable taxes from economic costs.
5 A shadow foreign exchange rate of US$1·27 per ₦1·00 as against the official rate of US$1·52 per ₦1·00.
6 Economic farm-gate prices for 1980 in 1974 constant Naira as projected by the World Bank's Economic Analysis and Projections Division, based on export parity for groundnuts and cotton and import substitution for maize, sorghum and cowpeas.
7 The exclusion of water and associated soil-conservation development costs, and project-evaluation costs, since the benefits attributed to them are intangible.

It is possible that the project might not meet the proposed production targets for a number of reasons, including poor management, lack of farmer response and the possible failure of delivery of farm inputs

TABLE 12.16 Nigeria: Funtua agricultural development project – Production levels as percentage of estimated project output

	100%	90%	80%	70%	60%	50%
Economic rate of return per cent						
Shadow foreign exchange rate	41	37	32	18	23	18
Actual foreign exchange rate	32	28	25	21	17	12

due to uncontrollable circumstances. For this reason it is important to test the economic rate of return against various levels of reduced production and the associated reduction in farm inputs, allowing all other factors to remain constant. This analysis is summarised in Table 12.16.

The project would have important secondary benefits that would accrue both to the people of the project area, and for the state and federation as a whole. The more important ones would be those listed below.

1 The people of the project area would benefit from:
(a) a general improvement in communications and public utilities, due to project actions initiating change and providing a basis for extending, by other government institutions, social improvements including health and education;
(b) an expected increase in local trade that would offer better opportunities of employment for the non-farming population; and
(c) the strengthening of local institutions including the local authorities and local department of government.
2 The North Central State would benefit principally through part of its population being stabilised and becoming more productive. Increased net employment accounting for 12 000 man-years per annum would result from projected actions. The state would be able to use the project to test new development ideas that could, if successful, be easily transposed together with trained staff to other areas. The Ministry of Agriculture and Natural Resources would be strengthened, and would derive from the project new credit and marketing techniques that could be easily applied to farmers in other states. The project would set up a seed-multiplication centre that would provide seed to other farmers elsewhere in the state. Finally, the project would produce additional raw material and would demand additional input resources to support the growth of the state's industrial expansion.
3 The Federal Republic of Nigeria would benefit from annual net incremental foreign exchange earnings amounting at full production to about ₦14 million (US$21 million). Project evaluation through the project-established evaluation unit would provide a wealth of information that could assist in determining federal agricultural policy, and resource and financial allocations throughout Nigeria.

Even though the economic rate of return appears very favourable, in the long run, indirect tangible benefits and intangible benefits may prove far more valuable than direct tangible benefits. Improved farm-management practices and profits to farming would far outlive the project. A step would be taken in the right direction to reduce income disparity, and to ensure a more even food supply for Nigeria's urban populations and expanding commercial livestock industries.

References

1 currency unit = Naira ₦
US$1.00 = ₦0.617
₦1.00 = US$1.62

13 Indonesia: seeds project

Introduction

This project is selected as a case study because of its location in one of the largest food-deficit countries in Asia. It is a technology-oriented project, which aims at the modernisation of Indonesian agriculture through the use of modern inputs – high-yielding seed varieties, in this particular case. The core of the problem was how to bring the fruits of genetic research undertaken by the International Rice Research Institute in the Philippines to the masses of Indonesia's rice growers while maintaining intact the genetic potential for high yields possessed by these varieties.

Briefly, by the end of the project preparation process, the objectives of the project were defined as follows:

> The project is the establishment of the initial element of a modern seeds industry in Indonesia, which, in a first phase, would develop a rice-seed industry aimed first at Java but would also lay the basis for a second phase which would cover other crops (e.g. maize and pulses) and other islands.

The major components would comprise: breeding and research facilities; seeds legislation, physical infrastructure including irrigation development, for seed production; equipment for mechanisation; seed processing and storage plants in selected villages; support for marketing of seed and distribution of inputs; technical assistance and training; and setting up of the required institutions for project implementation.

The wide spectrum of activities, ranging in this project from breeding through production and processing to marketing, requires careful planning of the organisational and marketing aspects. The reader will note that the identification process – possibly the most crucial in the project cycle – can itself be further broken down into rather distinct phases.

1 **The diagnostic phase** in which the existing situation is analysed at various levels, including the country, the agricultural sector, and the industry specific to the project, e.g. the rice industry in this case.
2 **The decision-making phase** in which the major question to be answered is whether the findings from the diagnostic phase warrant further detailed or feasibility studies being undertaken. That is, whether there is a project or not. The project's overall justification would then be established, e.g. its priority and general feasibility substantiated.
3 **The project-concept formulation phase** involves essentially two parts. First, there is the identification of key issues on which positions must be taken. Second, the project's concept – that is, objectives and broad

strategy – must be defined and project components determined. The determination of the project's scale, even very roughly, is part of this phase.
4 **The detailed project preparation phase**, in which the project's components are spelled out and the organisation and management features worked out. Costs, in particular, are worked out in detail and the financial, economic and social implications of the project are assessed.

The mission from the FAO/World Bank Cooperative Programme had the following terms of reference, at time of project identification.

1 Review the present status of the seeds industry in Indonesia and assess the government's plans for the development of the seed industry.
2 Evaluate the technical, legal and organisational aspects of the seed programme with special reference to research on plant breeding, varietal experiments, legislation on seed control, testing and certification.
3 Examine the existing arrangements and facilities for production of breeder seed, foundation seed and stock seed.
4 Examine the seed processing, marketing and distribution policies and arrangement, especially in respect of their consistency and adequacy.
5 Against the background of the foregoing, and the planned activities of other agencies, e.g. USAID and foundations, identify a project or projects suitable for World Bank financing.

The more important points arising from briefing discussions in Djakarta were that many donor countries and institutions were assisting Indonesia's seed industry but in an *ad hoc* manner because there were no clear indications or framework along which the seed industry should develop. An approach to the development of the seeds industry within which interested donors could make clearcut, rational contributions was therefore required. At the Sukamandi estate, very much under-utilised, in western Java, the government wished to gain experience in mechanised rice cultivation for application to the outer islands, and this should be considered in project formulation.

The more important aspects focused on in the field work were the existing seed programmes, including the work programmes and facilities of the main rice-breeding stations; the work of the Ministry's extension seed farms; arrangements for final multiplication of seed at the village level; distribution programmes and their operations for fertiliser and other inputs. Observations of the progress of the high-yielding varieties in farmers' fields received particular attention. The administrative structure and agricultural services at the provincial levels were closely observed. Sukamandi estate's potential for development was explored on the basis of available data. Rice cultivation and harvesting techniques were observed and farmers' production costs and returns derived. Programmes being undertaken with bilateral assistance from West Germany, Japan, the USA and the UNDP were assessed.

Apart from these specific observations, there are general impressions that every mission gets as it travels through a country. Rightly or wrongly, these somehow tend to impinge upon and influence one's attitudes towards the project.

The more important observations were: although Indonesia then ranked as one of the poorest countries in the world with a *per caput* income of about US$80, there were no clearly visible signs of abject poverty as one might see in comparable countries with similar *per caput* income levels. Although some areas in Java are very susceptible to periods of food deficit, there were no signs of emaciated pot-bellied children, or visible symptoms of kwashiorkor or acute malnutrition. However, frequent cases of avitaminosis – Vitamin A deficiency – leading to blindness were noted among adults.

Population pressure was strongly in evidence. The population density of 600 persons per km² for Java was readily believable as witnessed by the massive proportion of children in the populous villages and towns.[1] The intensity of the man/land pressure was clearly visible from the miniscule farm sizes prevalent in Java, which range about 0·25 to 0·5 ha. The labour utilisation intensity was seen in the rice-harvesting operation when hundreds of women harvested the crop from small farm units, by cutting individual panicles by knife.

The landscape itself was characterised by a sense of order. The well-terraced hillsides, with their cultivation on the contour, and controlled water flows, dropping from terrace to terrace over long stretches of mountains down to the valleys, provided a truly impressive picture of the need to conquer every possible bit of land. A further marked feature was the uniformity of crop stands at the village level, despite the fragmented farm holdings. The overall impression conveyed by the landscape was one of the 'hidden hand' at work, Hirschman's principle. This, in fact, reflected work of the village organisation coordinating planting times and water management.

Finally, some assessment of how a country maintains its stock of capital can be a useful indicator. Whereas in the post-colonial era, the large public irrigation schemes had suffered marked deterioration requiring extensive rehabilitation, the smaller privately operated schemes appeared well maintained.

The diagnostic phase

The assessment of the existing situation based on a preliminary evaluation of available data was designed to answer the basic question – 'Is there a case for a project or not?' The following review covers first the general setting or background, and the existing situation of the seeds industry as of 1969.

The general setting

Indonesia's population in 1970 was about 120 million with an annual growth of about 2·4 per cent. The archipelago includes some 13 000 islands of which 1 000 are inhabited. Ninety per cent of the land area is accounted for by Java, Sumatra, Kalimantan, Sulawesi and West Irian. Only about one-tenth of the total land area of 180 million ha is under cultivation, with the rest being mountainous or in forest. Java, with only 7 per cent of the land area supports over two-thirds of the population, some 85 million; its population density of 600 per square kilometre overwhelmingly exceeds that for the other islands, Sulawesi, Sumatra, Kalimantan, whose population densities range between 45 and 15.

Governmental efforts at reducing serious population pressures in Java by transmigration to the outlying islands have been, for various reasons, notably unsuccessful. Apart from the sheer magnitude of resources that would be required to make any visible impact on the problem, there appear to be many other factors, e.g. family, cultural and possibly religious, that make substantial emigration from Java a rather remote prospect.

With Java's land resources occupied to the hilt, it is axiomatic that yield-increasing innovations coupled with increases in irrigated land, provide the only real options with which the authorities can confront the situation.

Economically, Indonesia through most of the 1960s suffered excessive inflation, and determined government efforts had, by the late 1960s, in 1969, brought down inflation to an annual rate of about 85 per cent. The availability and price of rice, therefore, was a stabilising element, and government officials received rice allocations as part of their wages in kind.

The agricultural sector and the rice economy

The sector then accounted for about 55 per cent of national income and about 60 per cent of the total export earnings. Rice production dominated the agricultural economy, and accounted for about two-thirds of the total food supply with the *per caput* consumption averaging 90 to 95 kg annually. Java accounted for about 60 per cent of the rice-harvested area and about the same percentage of Indonesia's production.

The annual growth rates expressed as percentages, are shown in Table 13.1 for the harvested area, production and yield of rice. The data shows that rice-production growth rates of 2·0 per cent between 1953/54 and 1959/60 and of 1·1 per cent between 1959/60 and 1966/67, were

TABLE 13.1 Paddy rice – Annual growth rates for area harvested, production and yield (per cent)

	1953/54–1959/60	**1959/60–1966/67**
Area harvested		
Java	0·9	0·7
Outside Java	2·7	3·0
Indonesia	1·6	0·9
Production		
Java	1·5	0·8
Outside Java	2·8	3·2
Indonesia	2·0	1·1
Yield		
Java	0·6	0·1
Outside Java	0·1	0·2
Indonesia	0·4	0·2

steadily falling behind the population growth rate of 2·5 per cent annually.

Data from the Planning Board (*Bappenas*) and the Bureau of Statistics showed that the increases in production derived largely from new lands brought into cultivation outside of Java. The poor performance in Java arose particularly from two factors, namely, absence of new lands and a deteriorating irrigation network.

Since domestic production was not keeping pace with population growth, the gap had to be met by imports and by holding down actual *per caput* intake below the desirable levels. Thus, for the period 1968–1970, production of milled rice ranged from 10·4 to 11·1 million tonnes and imports averaged 0·63 million tonnes annually. At the then prevailing modest price of US$80 per ton, this still represented a substantial foreign-exchange drain which the government was determined to eliminate.

The rice economy then faced some major institutional problems. The first was that provincial governments effectively controlled the movements of rice outside of their provinces, with the result that lack of mobility contributed to imperfections in the operation of a national rice market. Government intervention in the rice marketing and pricing operation through the procurement agency, BULOG, resulted in depressed prices to farmers, with disincentive effects on production. These factors obviously constituted negative elements in the broad environment within which a seeds industry would have to operate.

The government's objectives

The five-year development plan, starting in 1969, aimed at self-sufficiency in production of food grains, and an increase in the *per caput* consumption of these grains from an average of about 1 980 calories to one of about 2 300 calories daily. The five-year cropping targets of the plan are shown in Table 13.2.

While much of the targeted increase in production for rice was planned to derive from new land to be brought into production outside Java, the largest increases in production were to be obtained by a considerable expansion of the rice-intensification programmes already operational for some years and administered by the central and provincial agricultural extension services.

TABLE 13.2 Target production for main food crops

	Rice		Maize		Pulses	
Year	million tonnes	per cent annual increase	million tonnes	per cent annual increase	million tonnes	per cent annual increase
1969/70	10·52	7·34	3·37	3·00	0·95	2·60
1970/71	11·43	8·65	3·51	4·25	0·99	4·21
1971/72	12·52	9·50	3·70	5·50	1·08	10·90
1972/73	13·81	10·30	3·94	6·50	1·21	12·03
1973/74	15·42	11·60	4·23	7·25	1·40	15·70

TABLE 13.3 Summary of rice intensification targets

Year	Anticipated total harvest area: million ha	Anticipated total harvest area: per cent annual increase	Rice intensification programmes: million ha	Rice intensification programmes: per cent of total harvested area
1969/70	7·60	1·33	2·59	34·1
1970/71	7·69	4·74	2·90	36·4
1971/72	8·32	4·52	3·15	37·9
1972/73	8·76	5·29	3·48	39·7
1973/74	9·30	6·16	4·00	43·0

The targets for the rice intensification programmes are shown in Table 13.3.

The rice-intensification programme would be undertaken mainly in Java, through variation of the BIMAS[2] programmes. This massive programme for distribution of inputs is discussed later, as this programme provided a major institutional consideration in the project's formulation. The nucleus of the intensification programme (*Pancha –u. Saha*: literally the five efforts) was a package programme comprising:

1 the use of high-yielding varieties;
2 the use of recommended fertiliser levels;
3 pest and disease control;
4 improved irrigation practices and water management; and
5 improved cultivation techniques.

The existing seed industry

Since the early 1950s, Indonesia had developed a formal system of seed multiplication in recognition of the benefits that could be derived from good seed. The predominant source of seed had been and continues to be 'farmer-saved seed', that is, seed reserved from the farmer's own grain crop. The system provided for all the stages in the progressive multiplication of seed.

The main elements comprised the breeding and testing activities of the Central Research Institute of Agriculture (CRIA). Its achievements then included the development of nine improved rice varieties and improved varieties of maize, groundnuts and soya beans. It had also tested and released two high-yielding rice varieties from the International Rice Research Institute in the Philippines. CRIA activities included the production of breeder and foundation seed.

Stock seed – the next multiplication stage – was produced by some 290 extension seed farms (average size 7 to 8 ha) scattered throughout all the provinces. Provincial governments owned and controlled these through their budgets, with the provincial agricultural extension services responsible for their operations.

Extension seed production – the last stage of multiplication, before release to farmers for growing as grain crops – was carried out on thousands of village seed farms under the theoretical supervision of the extension services. The farms, averaging about 1 ha each, were located on village communal land or private farmers' plots. Their total output was estimated at roughly 10 000 to 12 000 tonnes per year.

The foreign assistance programmes in seed – largely from the Japanese government and USAID – were geared mainly to reinforcing the government's proposals for development of the seed industry, through the rehabilitation and strengthening of activities at the provincial seed farms.

The government's more important seed improvement proposals for rice called for the following.

1 The provision of about 40 000 tonnes of seed to cover 1·3 million ha.
2 Selection and testing for distribution from among 3 000 rice strains and the rehabilitation of CRIA's sub-stations for production of breeder and foundation seed.
3 Establishment of six provincial seed centres for foundation seed production, seed inspection and testing.
4 Rehabilitation of 58 extension seed farms for stock-seed production.
5 The rehabilitation of Sukamandi estate into a model farm for mechanised production operations with rice-seed production as a part of the activities.
6 Equipping about 670 village seed farms for rice-seed production.
7 Various training programmes of a technical nature.

The government's proposed programme thus aimed at a consolidation and strenghthening of the existing seed industry structure.

Evaluation of the existing seed system

The analysis concentrated on answers to the following questions. Did the system meet the test of producing quality seed? Could the system cope with quantity? Was the system efficient in terms of resources, especially human resources?

The answer to the first question was no. At the earliest multiplication stages, the supply of foundation seed to the extension seed farms was unreliable both in quantity and quality. In turn, the flow from the extension seed farms to the village farms was interrupted and unreliable. Technical supervision of the village seed farms by the extension services was either minimal or absent, if only because the task itself was an impossible one because of its enormity.

The quality of seed can be gauged from inspection of the milled rice stock in markets. The samples observed showed variable sizes and types of grain and varietal admixture was widespread making it very difficult to produce a uniform milled product; there would be, therefore, a low milling out-turn which would increase costs. Moreover, the existing high seed-rates were themselves evidence of low percentage germination of the seed in general use. Farmers and agriculturalists were aware of the defective quality of seed in current use. It was concluded that the seed system was variety-oriented, and that farmers tended to associate good seed with seed of a particular desired variety.

One reason for the poor-quality seed stream was the inadequacy of budgetary funds both at the production and processing levels. The government-operated parts of the system – breeding, foundation and stock-seed production – were inadequately funded. In fact the provincial seed stations received no budgeted funds and were rather viewed as income sources for the provincial governments and also for the extension service staff. It was inevitable that quality should be sacrificed under such conditions.

The virtual absence of seed-processing facilities contributed markedly to poor quality. Inadequate drying facilities were further compounded by the absence of proper storage facilities.

The system's capacity to deliver quantity was quite elastic, though admittedly at an enormous cost to quality. To illustrate this, although the variety PB5 was first introduced only in 1966, seed supplies (by 1969/70) were available to cover up to one million hectares. The quality/cost element of the system was clearly visible in field stands of the varieties where significant varietal contamination was already occuring.

With regard to the efficiency of resource use, one problem was the massive amount of time spent by the extension services on seed-related activities. Procurement and distribution of seed for the various BIMAS programmes was considerably time-consuming. This, in turn, left the extension services very little time to do their substantive work of educating farmers and advancing the level of technology. Superimposed on all this was the absence of any seed law to protect the farming public, and of any national institution capable of implementing a national seeds programme.

The technical basis for a possible project

Rice varieties

A key issue in the adequacy of the technical basis for project formulation was whether there existed a stock of improved and genetically superior varieties, adequately tested, as a basis for launching a seed industry.

The CRIA was the agency responsible for rice-breeding work, with long experience in the field. Apart from the local stock of varieties, it introduced a strong flow of rice varieties and worked closely with the International Rice Research Institute (IRRI) in the Philippines. The existing varietal situation could be classified as follows.

Local traditional varieties These unimproved types occupied about 60 per cent of the planted area. They were favoured by farmers because of adaptability to various ecological conditions, reasonable disease resistance and, though low yielding and late maturing (140–170 days), they rated very high for their flavour and cooking qualities. Their long panicles with non-shattering characteristics made them well adapted for harvesting, transport and storage as stalk-paddy. They were well adapted to the higher-elevation irrigated areas, and also some lowland areas that were unsuitable for the improved varieties. Their low yield of 1·5 to 2·5 tonnes dry gabah (paddy) per hectare were a marked disadvantage.

National improved varieties The CRIA had developed nine such varieties, which were estimated to cover over 30 per cent of the planted rice area. They were much more responsive to fertiliser than the traditional varieties, but tended to lodge at threshold levels of nitrogen exceeding 60 units per ha. They had a long growing season (140–160 days) with reasonable disease resistance, and had acceptable flavour and cooking qualities. They were also adapted to existing harvesting, and

TABLE 13.4 List of rice varieties released by CRIA and their characteristics

Variety name	Year of release	Yielding capacity (tonnes/ha)	Disease reaction: bacterial blight	Disease reaction: bacterial leaf streak	Disease reaction: sheath blight	Disease reaction: blast	Maturity (days)	Elevation (m)	Cooking quality	Photoperiod sensitivity
Bengawan	1943	3·1	MS[1]	–	–	MS	155	500	good	indifferent
Sigadis	1953	3·3	R	R	–	HS[3]	145	500	bad	indifferent
Remadja	1954	3·6	–	–	–	HS	150	500	medium	indifferent
Djelita	1955	3·7	–	–	–	HS	160	500	medium	indifferent
Dara	1960	3·3	–	–	–	HS	140	500	good	indifferent
Syntha	1963	3·5	MS	MR	–	HS	148	500	good	indifferent
Dewitara	1963	3·4	–	–	–	HS	148	500	good	indifferent
Rathara	1965	3·7	–	–	–	–	148	500	good	indifferent
Arimbi	1965	3·8	–	–	–	–	150	500	good	indifferent
PB5 (IR5)	1968	7·0	MR[2]	S	S	S[4]	135	800	medium	slightly sensitive
PB8 (IR8)	1968	7·0	S	MS	MS	S	125	800	bad	indifferent

[1] MS = moderately susceptible
[2] MR = moderately resistant
[3] HS = highly susceptible
[4] S = susceptible

transport and storage methods. Their yield potential varied between 2·5 to 4·0 tonnes per ha depending on input levels, but their impact on total production was quite large because they occupied the bulk of the irrigated areas with a potential of two crops per year.

High-yielding IRRI varieties The demonstrated yield superiority of two varieties from the Philippines, renamed PB5 and PB8, over all local improved and traditional varieties was beyond question. Yields of 4–6 tonnes per ha were commonly reported. The yield potential, however, up to 7·0 tonnes per ha in experimental trials (see Table 13.4 and Fig. 13.1) was only attainable under conditions of controlled water supply – that is, reliable irrigation and with considerably higher dosage levels of fertiliser than for the existing varieties. The thicker straw and shorter length of these varieties enabled them to respond to higher fertiliser input levels without lodging or collapsing like the traditional varieties. The varieties, introduced in 1966, were, however, not adequately tested under varying ecological conditions and cultural practices before their widespread release. It may be argued that with the desperate conditions of food deficit and inflation then facing Indonesia, the country had no real option besides undertaking a crash programme for the widespread multiplication and distribution of these varieties which opened new vistas for rapid production increases. This is precisely what the country did, but the inadequate adaptive testing work left its toll.

For example, both varieties proved susceptible to bacterial blight, especially PB8, which later had to be withdrawn. The other defects were

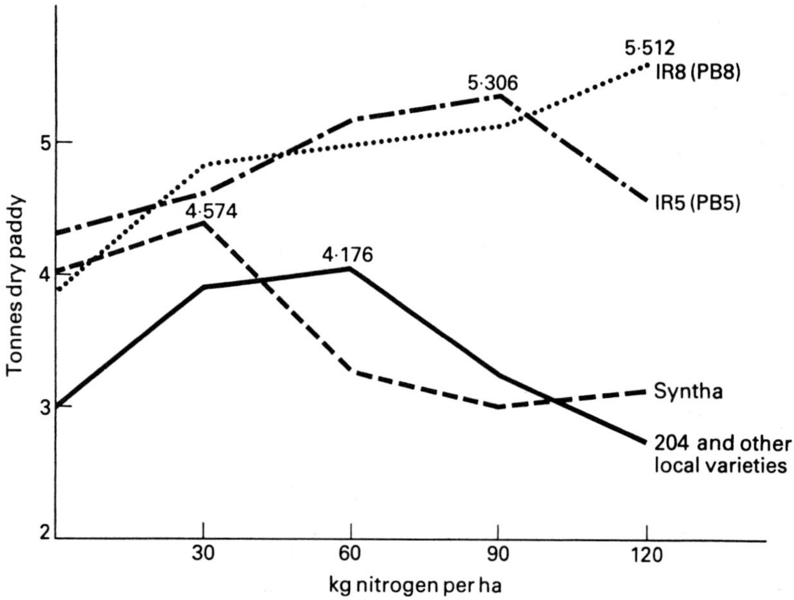

*Source: Central Research Institute for Agriculture, Bogor

Figure 13.1
Indonesia: Seeds — Varietal response to fertilisers.

that the panicles shattered readily, requiring growers to modify their traditional systems of harvesting, transport and storage. Organoleptically they were far less acceptable than the traditional varieties, with their flavour being rated low because of their 'mealy' characteristics. The market thus sharply discounted this and prices were some 30 per cent below those paid for the best cooking varieties.

The wide and rapid spread of the varieties among neighbouring farmers did, however, have an impact on production. The deficiencies of the existing seed-multiplication system resulted in the fact that only three years after their introduction, these varieties were already showing signs of deterioration and contamination by other varieties.

Varieties in process of development The CRIA, with assistance from IRRI was studying a large number of newly developed lines and introductions, some of which were soon to be ready for release. These varieties, e.g. C4–63 had much better flavour and cooking quality than PB5 and PB8. The stage was set, therefore, for a pipeline of new varieties. This technical work at the breeding level was reinforced at farm level by the fact that the Indonesian rice-grower was quite efficient, and responsive to new techniques considered to be economic.

Institutional considerations The more important considerations involved BIMAS and INMAS programmes. These are essentially farm-input package programmes. The BIMAS programme arose as an extension device. The programme which began in 1963 was based on demonstrations from 100 ha at the Agriculture College at Bogor, which showed the large increase in rice yields possible through the use of improved seeds, fertilisers, pesticides, improved practices and better water management. Selected blocks of farms in the more favourable farming districts formed the initial basis for the programmes, where farmers received a package of inputs in kind for repayment in kind as a share of the crop. By 1967 the system was changed with the granting of cash loans and repayment in cash. Under the BIMAS programmes the extension services provided technical guidance to farmers and small groups of farmers received collective credit from the Bank Rakjat Indonesia (BRI). After two or three years with the BIMAS programme, the better farmers were expected to graduate to the point where they could purchase these inputs with their own resources, thus entering the INMAS programmes. By 1969/70, the BIMAS programme covered about 840 000 ha and INMAS about 330 000 ha.

Several variations were introduced into the BIMAS programme as they evolved. BIMAS *baru*, introduced in 1967, emphasised the use of high-yielding varieties in the package. BIMAS *gotong royong* introduced a completely new concept. Large foreign companies contracted with the government on delayed-payment terms to supply fertilisers and insecticides to farmers in large contiguous blocks each of about 100 000 ha, largely in Java. The companies not only arranged the logistics for moving these inputs to the designated areas but also undertook the aerial spraying of paddy fields with insecticides where possible. Farmers participating were required to repay with one-sixth of their crops to BULOG which was responsible for payment to the foreign fertiliser companies. Farmers within the selected areas were obliged to accept the package with a value of up to US$60 per ha.

The latter programme ran into problems, because of the substantial losses incurred by BULOG. A later change involved the replacement of the foreign-based companies by a domestic operated programme – BIMAS *nasional* involving a credit programme administered by BRI, with a national policy board. Although these various programmes cannot be fully assessed here, some weaknesses were clear. For example, the recommended packages tended to be rather inflexible and were arrived at without adequate consideration of the varying requirements of different selected areas. However, it was clear also that the BIMAS programme contributed markedly to yield increases in the selected areas, and that the programme would continue in some form or another. Equally obvious was the fact that the quality and quantity of seed required by the programme were not being supplied by the existing formal seed streams.

The Ministry of Agriculture

The directorate of particular interest was agriculture with a director-general controlling five technical divisions and three agricultural research institutes, one each for agriculture, horticulture and soils. However, 26 provincial agricultural services set the policies. The directors of these services were administratively dependent on the provincial administration but technically responsible to Djakarta. A serious weakness of these services was the ratio of extension workers to farmers. In Java, for example, with 6 900 extension workers, every extensionist had to serve about 1 600 farmers. The low qualifications of the extension staff, a serious problem by itself, were compounded by their highly unsatisfactory working conditions. Transport facilities were absent or hopelessly inadequate, and the budgets were almost entirely devoted to wages and salaries with hardly anything for equipment. As previously noted, the duties of the extension staff were varied and burdensome.

The Bank Rakjat Indonesia (BRI)

The BRI's main operations were geared to financing agriculture, fisheries and cooperatives. Though capitalised at US$0·8 million, only about one-third had been actually paid in. Its resources comprised loans from the Central Bank (about US$71 million) and clients' deposits (about US$45 million). It had recently built up a network of branches throughout the country and operated 13 regional offices, 206 branches, 13 sub-branches, 81 village units and 400 mobile units. The bulk of its loan operations involved short-term financing largely for rice, of which 20 per cent were for BIMAS programmes. Low repayment rates meant that a large volume of bad debts had accumulated. The collection rate averaged 50–60 per cent of the BIMAS credits. The BRI had operated mainly through village cooperatives but as this had proved unsatisfactory, it was developing its own network of village units for serving individuals directly, and thus permitting more intensive supervision.

Rice marketing and pricing policies

Existing rice-marketing practices and the government's pricing policies constituted disincentives at the farm level for the rice producer. The absence of a national rice market was a problem. Trading of rice and paddy between provinces, and sometimes even at the *kabupaten* level within the same province, was subject to restrictions by provincial laws. The provincial administrations could also suspend the procurement activities of BULOG within their provinces. Because the market could not readily match surpluses and requirements of deficit areas, there were wide price variations between regions, and this sometimes resulted in higher procurement costs to BULOG than if normal marketing practices prevailed.

The price difference between regions bore little relationship to normal transport and marketing costs. Furthermore, since only an estimated 20 to 30 per cent of total rice production ended up in commercial marketing channels, small changes in total production tended to induce disproportionately wide price changes over the same period. This, coupled with the absence of a proper price-stabilisation programme, induced not only wide price disparities between regions but also wide price fluctuations at various seasons of the year. Governmental rice-price policy aimed at maintaining a relatively low domestic price for internal consumption. BULOG's operations provided the main thrust through its procurement operations, both domestic and foreign. In 1969 BULOG's targeted distribution of rice in tonnes for that year was:

Government officials – civilian and military: 655 000
Pn/PNP (State Enterprises): 240 000
Free market sales: 120 000

BULOG's foreign operations had an adverse impact on rice traded by domestic producers. The rice prices prevailing were then quite low. The large amount of wheat flour imports and the pricing policy relating to wheat flour, especially PL 480 imports, accounted largely for this. The price of wheat flour was fixed at a level of about two-thirds of the rice price. The retail wheat flour price had settled at between Rupiahs 36 and Rupiahs 38 per kg. Prior to 1968, wheat flour prices were always higher than milled rice prices, but in 1968 BULOG began the distribution of wheat flour through normal commercial channels, and heavier wheat imports were planned in 1969.

A conscious effort was also being made to establish some working relationship between the price of rice and key inputs. At the time of project formulation the rice price/urea ratio was moving rapidly against the farmer. The later introduction (1969/70) of a price-support policy was intended to give farmers greater incentives to use fertilisers and other inputs.

The decision-making phase

On the basis of the analysis of the situation summarised above, the decision had to be taken whether or not to commit substantial additional funds into Indonesia's seed industry – that is, whether a case existed for a seeds project or not.

The main considerations favouring a positive stand were as listed below.

1 At the national level, Indonesia was a country with a food deficit, with food imports, especially rice, constituting a serious drain on foreign exchange. A seeds project could be justified within the national context, if it could make a notable contribution to the government's objectives, expressed in its five-year development plan, of substantial production increases of food grains especially rice, the staple food of the diet. Thus a seeds project would meet the first requirement – that it had high priority in resource allocation to meet justifiable national objectives.

2 The contribution that a seed project could make to the modernisation of Indonesia's agriculture was a weighty factor to consider. Modernised agriculture always entails a higher level of purchased farm inputs than a traditional system. The evidence showed that the quality of seed at the farm level was definitely a constraint to the larger use of other purchased inputs, especially fertilisers, since by far the larger bulk of the rice-plant population was genetically incapable of utilising high levels of fertiliser inputs – thereby posing an economic constraint on the expanded use of inputs. A quality seed stream therefore had a pivotal role to play in any plans for production increases, especially in Java, where sizeable irrigation investments were in process and planned.

3 The existing seed industry had failed to provide quality seed for the farmers. Whereas the system was capable of rapid bulking up or multiplication of desired varieties, it did this at the cost of a rapid deterioration in quality of the nation's seed stock. This was inevitable because of the sheer physical impossibility of proper administration and supervision of the existing system.

4 The substantial but fragmented external aid in seeds could not exert an impact because of the absence of a strategy for the seed industry's future development. A properly conceived seeds project would provide such a framework.

5 A technical basis existed for the launching of a seeds project. Though rather limited, and inadequately tested, the introduced IRRI varieties, both those already released by CRIA and those in the pipeline, had demonstrated their superior yield potential under Indonesia's conditions. They could, therefore, with the improved local varieties, constitute the core of a varietal package for feeding into a new seeds stream. This was reinforced by the fact that farmers were willing to pay higher prices for seed of the desired varieties, implying some awareness of the importance of good seed among farmers.

6 The institutional basis, though weak in parts could, with modifications and improvements, provide adequate support to ensure the successful prospects of a seed industry. The BIMAS programmes provided a secure outlet for a possible seeds project, and had shown their capacity to deliver massive quantities of fertilisers and pesticides at the village and farm level, with minimal facilities for intermediate storage. Thus, seed could also be made highly mobile.

7 With regard to support services, the credit facilities of the project would be modest and not impose any undue burden on the existing banking and credit system. The extension services, though heavily involved in activities related to the seed industry, especially procurement, had a vested interest in the existing seed system – a factor to be reckoned with in any new project proposal. At the village level the tight

organisational discipline built around the *lurah* – the village head – was a major positive factor. However, possibly the most negative element in the institutional setting was the land-tenure system as represented by the dominance of very small farm sizes in Java, because this posed severe organisational problems in the design and implementation of a seed project of any reasonable scale. Despite these constraints, the case for a project was justifiable.

The project-concept formulation phase

The following principles, based on experience of countries with established seed industries, were taken into full account in the formulation of the project's concept, in addition to local factors.

1 Seed processing and seed-producing activities must be closely integrated to ensure a minimal guaranteed amount of seed for processing. This is especially important since seed-processing facilities have little or no alternative use, and seeds-processing investment can account for a significant share of total project costs and foreign exchange requirements. Most processing plants for agricultural products, e.g. rubber, sugar cane, tomatoes, dairy, etc. require a raw-product supply within a certain radius from the plant, if the overall operation is to be economic. The same applies to seed, especially of grain crops.

2 A corollary of the first is that seed production must be concentrated in seed-growing districts. This would ensure the production of quality seed by facilitating concentrated guidance by the extension services, by allowing intensive supervision by staff specially trained in seed production, and by facilitating the movement of farm inputs into the area and products out of the area.

3 All investment at the farm level should be geared to ensuring a reliability of supply. A reliable water supply must be provided for in the investment-mix either through surface irrigation, pumping from streams or underground water supplies. Other farm investment may be required, such as land levelling to permit the required crop rotation and year-round production.

4 Grower–processor relationships should be such as to ensure that growers specialise in and remain in the seed-growing business. This calls for a system of incentives whereby the financial returns to the specialised seed growers are such as to guarantee that they will supply the processing plant with their seed crops. This, in turn, requires contractual arrangements, including a pricing formula for the seed crops that establishes some relationship with the grain price.

5 A seed project typically has or could have three major groups of activities quite different in nature: (a) the commercial aspects including seed multiplication, processing, and storage, marketing and distribution; (b) the regulatory aspects covering seed certification and the administration of seed laws; (c) the support services, such as research and plant breeding and varietal testing; extension, credit administration, etc. These various aspects must somehow be integrated into one working model.

6 The whole production system should be so designed that it is flexible with respect to output and demand requirements, since demand forecasting is a difficult exercise at best, with demand partly determined by the flow of new varieties into the system.
7 Whereas the economic justification of a soundly conceived seed project can be taken for granted because of the enormous multiplier effects along the line, the financial implications, particularly at the processor's level, must be closely evaluated and financial solvency guaranteed by the project. The injection of highly capital-intensive components, as represented by sophisticated seed-processing plants, into a static system geared to low-value crops – in contrast to more valuable products such as vegetable seeds – does pose financing problems especially during the period of building up a market. The budgetary implications of this would have to be closely studied.

An evaluation of the Indonesian situation against the background of India where a similar project had been previously formulated showed that in India there were distinct advantages. For example, the technical basis was much more firmly established. The work of the Uttar Pradesh Agricultural University on breeding, varietal testing and adaptation, was much more advanced than the work of the CRIA. The University had experience in seed production and processing. The project area in India was characterised by fairly large farm sizes, which in contrast to Indonesia with its very small units, simplified the task of organising production. At the national level, the institutions in India were well developed, with a National Seeds Corporation and an emerging system of seed laws providing a regulatory framework. No such institutions existed in Indonesia, thus making project formulation a more difficult task.

Main project issues – a discussion

Major issues in the project must first be identified before the project's components can be assembled. The broad parameters are roughly synonymous with what we might call 'the project concept'. It was indicated earlier that the project-formulation process essentially comprises the progressive resolution of key issues. This process is illustrated below.

The first issue was: Should a seeds project in Indonesia aim at strengthening or building around the existing seed system or should the project aim at establishing a completely new seed stream? The reader will recall that the main elements of the existing system comprised a pyramidal structure with the CRIA at the top with its breeding and foundation seed production; the provincial seed stations scattered throughout the country producing the next, stock-seed stage; and finally, at the village level, thousands of village seed-farms doing the final multiplication into extension seed which ends up with the farmer for production of grain. The basic principle implicit in this system was that the final product – extension seed – should be produced as close as possible to the point of end-use, thus requiring the involvement in seed production of as many villages as possible.

This principle was in direct conflict with the earlier stated principle that seed production should be concentrated in a limited number of carefully selected seed-growing districts. There was just no practical way of introducing into the existing system, the required seed processing and storage facilities; the necessary degree of field supervision and technical guidance for credit administration to seed growers, and arrangements for transport of inputs and product. Thus the primary project objective of quality, reliability, and quantity of seed supply could not be attained under the existing seed system.

The involvement of the CRIA within a new project by project actions designed to strengthen it was quite feasible, but there were grave doubts about the extent to which the provincial seed stations should retain their function of stock-seed production in any new project. Thus, for all practical purposes it was necessary to think in terms of a completely new seed stream and build new mechanisms for this end. Planning, therefore, called for the progressive displacement of the old seed stream by the new seed stream, fully recognising that entrenched difficulties would have to be overcome. The basic planning premise for project formulation, therefore, was that the project must ensure adequate control over all the stages of the seed-production process, from breeding through foundation seed, through stock and extension seed production and processing through to distribution.

The second issue was: What should be the geographic coverage of a new project? Obviously, there was just no practical way to design an initial seeds project covering the vast and widely spread islands of Indonesia. The organisational and logistical problems would be too much to cope with adequately in an early stage. It was decided, therefore, to limit the project activities to the islands of Java and Bali only, so as to facilitate project implementation.

A third issue was: What crops should be covered in an initial project? The government's food-production targets aimed to increase not only rice, but also maize and pulses. The CRIA had a maize-breeding programme focused on the development of high-yielding synthetic, composite and open-pollinated varieties, and had developed and distributed four varieties aimed at the main maize growing areas. Some improved varieties of soya and groundnuts had also been released. However, they had all deteriorated badly, and showed urgent need for repurification. There was some merit in the inclusion of other crops, e.g. sorghum, oil seed crops and cassava in any national plant-breeding programme.

However, the primary concern in a first phase should be to develop a workable system of seed production. If the mechanism worked satisfactorily for rice, it could later, with some adjustment accommodate other crops, e.g. maize, pulses, etc. Limiting the product stream to rice only, would simplify the problem of implementation during the stage involving problems of institutional building, market creation for quality seed, and training of staff and farmers.

A fourth issue was: How far should the project go in the provision of basic supporting services to a seed-production system? This led squarely to the question of precisely what the objectives of the seeds project should be since objectives determine project components. Should the project be limited to breeder seed from CRIA and multiplying this through a seed-production system that the project would establish? Or should it be concerned with the much broader issue of influencing CRIA's own pipeline? It was reasoned that only a very limited impact with a seed-multiplication and processing investment could be made by the project without addressing itself to the question of what determined the quality and composition of the nation's seed stock.

Stated simply, this seed project would have to cater for a national crop-improvement programme which should 'strengthen Indonesia's capacity first to pure-line its improved crop varieties prior to their inclusion in the seed-production system and second to originate new improved varieties'. This approach was adopted for several reasons. It was necessary to guarantee a continuing stream of new varieties, so as to stimulate demand at the farm level, and help ensure the financial viability of the seed production and seed-processing enterprises. The varieties released should, as much as possible, meet the varying ecological requirements of different parts of the country. The project, therefore, would have to provide the required investments to ensure a plant breeding and varietal adaptation and testing programme to sustain the continued pipeline requirements for a viable seeds project.

The fifth issue was: What role should the Sukamandi estate play in the seed project? Although this 4 000-ha estate represented a tremendous asset in Java's land-hungry situation, its introduction in the project posed problems. First, it would imply making rapid changes from an established small peasant-operated system of seed multiplication to a large-estate approach to seed production. The scale of operations would require mechanised cultivation and since there was no experience of large-scale mechanised rice farming in Asia, this would certainly raise many technical issues.

However, a major objective of the government's planned agricultural development of this estate was to develop practices for mechanised rice cultivation to be applied, not in Java, but in the out-lying islands, and to this end the USAID had already introduced about a quarter of a million US dollars of mechanical equipment. The injection into the project of the appropriate technical assistance and training of nationals would help to cope with the problem. Although Java did not suffer from a labour shortage as such, yet it would not have been feasible to cultivate such a large hectarage – up to 1 500 ha – without some mechanisation, since seasonal labour shortages could be expected during peak periods of transplanting, land preparation and harvesting.

Apart from the mechanisation aspects of the problem, a major area of concern was that this would substantially involve the government or the public sector in the commercial activities of seed production with the attendant problems of management. There had to be some sort of balance between the public and private sectors, and only a part of the estate would be used for seed production.

The sixth issue was: How to organise the seed activities in the other areas of Java? A reasonable proposal was that each of the three provinces in Java plus Bali, should have its own seed-multiplication system so that the seed requirements of the provinces would be basically guaranteed and that transport costs would be kept to a minimum. This would also minimise the possible risk of confrontation with the provincial authorities by not having to move large quantities of seed across the border.

An indication of how to organise this was provided by the satisfactory operation of tobacco-growing operations under contract, involving very small farm sizes, and farmers' compliance with requirements of the tobacco drying and packing enterprises. Contract growing had also been satisfactorily introduced for maize by the Japanese in eastern Java. It was logical, therefore, to plan for seed-growing districts, where farmers would be grouped together and would specialise in the seed-growing business, so as to meet the requirements of the seed-processing plants to be established nearby. The more important criteria in selecting seed-growing districts included the absence of flood hazards; proper road and/or rail services; and reliable irrigation facilities to permit double cropping.

The proposed new system could, however, only work if strong incentives were offered to induce potential seed growers to enter and remain permanently in the business. The package would have to include required on-farm improvements, e.g. improved irrigation and drainage networks, necessary credit facilities, good technical supervision, all geared to ensure that the financial returns of seed growers would exceed what could be obtained from the alternative use of his land and labour resources. Furthermore, exhaustive field checks were undertaken to establish whether the system would operate in practice. Apart from discussion with the staff of the extension services, the village heads, *lurahs*, in potential seed-growing areas were interviewed. A highly effective power structure prevailed whereby, the *lurah* was able to mobilise the villagers to undertake an agreed project that was in the interest of the community.

The seventh issue was how to set the scale of the project? This is particularly difficult in a seed project, where estimating the farm-level demand for the project seed involves much qualitative judgement about farmers' attitudes, motivations and receptivity. There are further considerations in arriving at project scale. The directly productive investments, seed production and seed processing, set the physical output limits of the project, and determine the scale of production. Thus, once production targets are set, cost estimates can be derived. However, supporting activities, e.g. research, training and extension services, can be substantial cost elements with little or no direct or immediate impact on the physical output targets. These costs of support services represented about 22 per cent of total project costs, estimated at time of identification.

The first target proposed was 40 000 tonnes of certified rice seed annually at full development, which at a seed rate of 40 kg/ha would permit the planting of 1 million ha per year. Some 2·4 million ha of land would be covered, for which a cropping intensity of 125 per cent was assumed, thus giving 3 million ha of ground planted to either fresh certified seed, or to first or second generation seed saved from certified plant plantings.

At the preparation stage discussed below, a consultant team scaled down the seed-rate assumption from 40 kg/ha to 30 kg/ha and considered a demand figure of 30 000–35 000 tonnes annually as a more realistic estimate, while at appraisal they were further reduced to a range of 20 000–25 000 tonnes annually. The considerable scaling down of the demand estimates arose primarily from changes in demand estimates of the various BIMAS and INMAS programmes.

The more important coefficients in arriving at demand were: the targeted area; assumptions as to frequency of seed replacement and the seed-rate per ha for final planting. In 1969 Java had a rice-harvested area of about 4·25 million ha of which 4·0 million were of lowland *sawah* rice. The project should aim initially to concentrate on the BIMAS and INMAS programmes, which had been selected for their potential for intensification. These technically irrigated and semi-technically irrigated areas were estimated to rise to 2·5 million ha by 1976 as compared to 1·6 million ha in 1970. These better-irrigated areas were the ones where government proposed to strengthen farmer-support services, including extension, credit and input supplies.

The frequency of seed replacement by farmers, i.e. the time period elapsing before a farmer purchases a new lot of seed, under normal conditions would be determined by the cost of seed and the benefits he expects from it. It was necessary, therefore, to work out end-user benefits for such certified seed using certain pricing assumptions. For the BIMAS/INMAS areas a seed-replacement frequency of 2·5 years was assumed. Finally, a seed-rate of 25 kg/ha was assumed for the new seed stream, since farmers who purchased seed from known sources used this seed-rate. It was also recognised that demand could be affected by the introduction of market promotion activities.

The project components

The project components at the identification stage, comprised the following:

1 The provision of physical infrastructure to the Sukamandi estate including irrigation, land levelling, a road network, domestic water services and electrical power, and processing plant facilities for the production of stock and extension seed.
2 The establishment of a National Crop Improvement Centre (at Sukamandi) to purify the existing stock of varieties and to originate new improved varieties. This would involve providing laboratory facilities, housing and offices.
3 Establishing a National Seeds Centre (at Sukamandi) to undertake seed analysis and grow-on trials, in association with branch provincial laboratories that would be established, and the training of personnel for seed extension, seed certification, etc. Physical facilities and dormitories and required buildings would be provided.
4 The provision of equipment and field machinery to mechanise production operations on about 2 000 ha (half of the estate), with required buildings.
5 The development of 12 seed-growing districts in the rest of Java (with a minimum of 500 ha each), and a small unit in Bali. Each district would have five seed villages, growing a minimum of 100 ha each. These would be selected on the basis of criteria earlier discussed.
6 Establishment of seed-processing and storage plants at Sukamandi and

in the selected seed-growing districts. Location should be such that no village is more than 10–15 km from the plant.
7 The establishment of the required institution to manage the project including the commercial and non-commercial activities (discussed under organisation and management).
8 The provision of technical assistance in establishing the project and training Indonesian nationals.
9 Assistance to the agricultural extension services with transport and possible salary supplements and additional work aids, e.g. films, posters.

The project-preparation phase

After agreement had been reached with government officials on the main proposals of the identification mission, consultants were engaged to undertake a study in accordance with the model seed stream and the institutional framework set forth at identification, and to prepare a detailed project proposal suitable for financing. Consultant services were deemed necessary because of the specialised skills required for seed processing and for the detailed engineering of the Sukamandi estate. The consultants' detailed terms of reference covered the following areas: seed requirements; research proposals; seed legislation and control; seed production; storage and processing; commercialisation; transportation; staffing and training; and technical assistance.

A summary outline of the chapters required was provided to the consultants. For example, the chapter dealing specifically with the seeds-investment project should include:

1 the project definition;
2 major project works and actions;
3 other project works and ancillary services;
4 design specifications of proposed facilities and equipment;
5 the project schedule;
6 cost estimates, including foreign exchange components.

The consultants' findings closely corroborated the findings of the identification mission. Their organisational proposals were broadly similar, and their institutional proposals included a national seeds-law, specific rules and regulations for the enforcement of this law, and specific regulations for producing foundation, stock and extension rice seed. They further selected specific sites for seed-growing districts, and for location of seed-processing plants with the required designs and costings.

Breeding and research Project actions were designed to ensure that the CRIA was equipped to accelerate the flow of new varieties of rice, and later maize, into the project's pipeline and to supply the foundation seed required. The project would finance technical expertise in rice and maize breeding for the CRIA in addition to required equipment. The CRIA also would be provided at Sukamandi estate with land and research facilities for this work.

Seed regulation The project would establish a Seed Regulation

Directorate to implement the legislation to be developed. For seeds-testing work, three provincial laboratories would be established. Since the project aimed to provide 'certified seed', the project would have to staff such a service and provide a seed laboratory and transport facilities.

Equipment of Sukamandi estate The project actions called for development of about 2 000 ha of the estate with irrigation, roads and land-shaping. Additional field and processing plant machinery would be provided for the production of almost all the stock-seed requirements of the project and about 8 640 tonnes extension seed annually. A 50-ha block would be developed and equipped with facilities for the CRIA to undertake testing and research activities. The rest of the estate, not used for seed production, would receive some irrigation facilities and a skeletal road network, and would be rented out to farmers for growing rice for consumption. With respect to irrigation development, the estate fell within the command area of the Djatiluhur dam, and an existing canal supplied water to a control structure at the estate boundary. On the mechanised farm, tertiary canals, together with associated canal structures would be designed for an irrigation duty of 20 litres/second per ha, sufficient to cope with the maximum requirements of the cropping pattern. The latter envisaged a double cropping of rice with an intensity of about 150 per cent to permit a rotating fallow.

The projected road-development would consist of rehabilitating the existing road network and construction of bridges. Proposals for land development and mechanisation took account of the inadequacy of labour in the area. Furthermore, if timeliness of operations was to be assured, the heavy soil types and the growing of two crops per year required mechanisation. Provision was thus made for plant and equipment for cultivation and harvest operations and land levelling of 2 000 ha.

Seed processing and storage facilities were also included at Sukamandi. A single standardised plant module was designed for installation at all sites with emphasis on flexibility, with respect to output levels, and number of varieties to be handled. Facilities would include a pre-cleaner, column drier, fine cleaners, seed dressing and bagging facilities, and conditioning and storage silos. The planning called for the processing plants to store the processed seed until prior to the peak planting season's requirements when they would be released to the major distributors.

District seed-processing plants and service areas A modular concept was developed whereby a module consists of a processing plant with a service or seed-producing area of about 500 ha cropped at about 150 per cent intensity, and each area distributed among 8 to 12 villages. The objective was to get groups of farmers (with their small farm-sizes of 0·3 to 0·5 ha) to work in discrete blocks of about 10 ha. This was basic because of sanitation requirements, so that varieties could be isolated; for economics of transport and to facilitate intensive supervision. The key to success lay in mobilising the village organisation structure through the *lurah*, and the provision of incentives. The proposal was that the project would absorb the threshing and transport costs and would pay seed growers a price premium of about 25 per cent above the prevailing price of standard wet stalk-paddy in the project area. Farmers would, in addition to their required inputs package, receive training in seed-production techniques.

Market development and distribution The success of the new seed stream would depend on several factors, among which were the rate at which demand could be created for the product, which in turn hinged strongly on the extent to which farmers' attitudes could be influenced. also the ready availability of the seed. A sales-promotion budget was therefore developed, to be used for developing sales staff, for demonstration programmes by extension service, and to reinforce the facilities in the proposed distribution channels.

Training Provision was made for training local personnel in seed certification, seed analysis and seed production, and in processing. Local in-service training would be combined with overseas training especially for the processing plant managers.

Technical assistance Provision was made for experts in seed-industry administration, seed certification and testing and mechanisation, and in seed-processing management to assist the project. Additionally, the CRIA would receive support in its breeding programmes for rice and maize.

National Seed Board

Overall responsibility for the project would be vested in the Ministry of Agriculture. Figure 13.2 shows the proposed organisation structure. The Board's main functions would be to advise the Minister of Agriculture on the planning and formulation of basic government policies for the development and operation of a national seeds industry. Such policy issues might comprise pricing of seeds, varietal releases, control of seed imports, the role of private enterprise, and the phased development of the industry. The proposed composition was broadly to represent a wide range of pertinent interests. Each institution has to have a legal basis – in this case it was confirmed that a Presidential Decree establishing such a board would suffice. It must be noted that the Board would have only staff authority and no line authority.

The Seeds-regulation Directorate

The major functions would comprise: seeds-testing involving testing field samples at all levels, e.g. breeder seed through to extension seed, and seeds-certification to control and warrant all seed bearing the Directorate's certification labels, through field and processing plant inspection.

The location of this service in the organisational hierarchy was very important because of the strict independence it must have in its role as watchdog for product quality. It should have powers to reject unsatisfactory batches of seed that may be produced by the National Seeds Corporation. Because of this powerful control function, its location was proposed outside of the seed-production stream with direct responsibility to the Minister of Agriculture (Figure 13.2). The legal basis for establishing such a Directorate would be a decree by the Minister of Agriculture. The directorate would charge fees for its services.

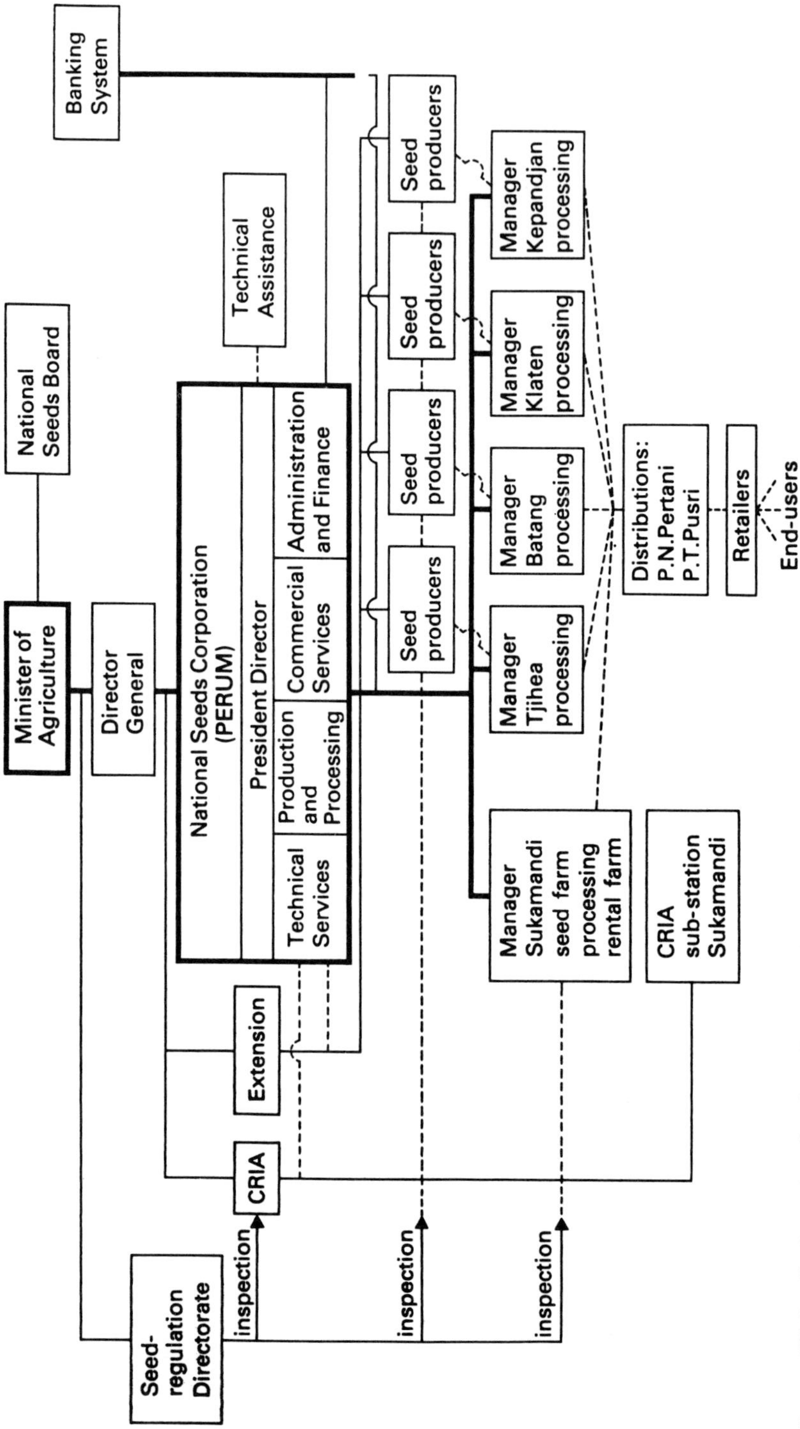

Figure 13.2 Indonesia: Seeds – Proposed organisation

The National Seeds Corporation (NSC)

The NSC would be responsible for all the commercial activities of the project, including seed production, seed processing and storage, and marketing.

The choice of corporate structure required careful consideration and discussion with legal officers of the government. There were two important factors – first, while the project's economic viability was undisputed the financial profitability was very marginal and difficult to forecast. While costs could be determined within a reasonable margin of error, the gross and net returns would be determined by volume and price of seed sales, both prone to wide error. Because of this, it was imperative to plan for a structure that could accommodate a government subsidy, if this was required. The other consideration was the possibility, later rejected after discussion with government, to accommodate private foreign enterprises, especially in the processing and marketing aspects of the project. An international seed firm was then very interested in participation in the project, but the government decided that the project should be managed entirely by the public sector at the initial stage. The option selected would not have accommodated private enterprise.

The NSC was to be established by government regulation as a wholly state-owned enterprise or *Perusahaan Negara Umum* (PERUM, for short), within the framework of Law No. 19 of 1960 governing state enterprises. A state enterprise is a legal entity, having its own capital separated from the state treasury and, as such, is an independent economic unit which is engaged in public utilities, services and other profitable undertakings. It employs cost-accounting principles and is able to contract loans, both foreign and domestic.

The PERUM has a management structure which must follow a legally prescribed form, which would permit closer and more direct linkage to be developed between the commercial and non-commercial activities of the project. As seen in Figure 13.2, the Directorate General of Agriculture would be able to coordinate the PERUM's activities with those of the CRIA and the extension services, for example. The financial structure of a PERUM also makes it possible for government to provide annual budgetary contributions for its operations which would be necessary in NSC's operation in the early years.

In contrast, the other option, *Perusahaan Negara Perseoran* (PERSERO, for short), is defined as a state enterprise constituting a Limited Company based on the commercial code. This would have permitted a joint venture operation with, say, foreign private enterprise. The government's wish to keep the project wholly within the public sector initially, made the use of a PERSERO impractical. However, the legal framework was such that a PERUM could later be converted into a PERSERO.

The functions, proposed for the NSC included the following.

1 The management and operation of the 2 000-ha seed farm at Sukamandi, and also the running of the non-seed farm activities, that is, the rental farm at Sukamandi.
2 The establishment, operation and maintenance of seed-processing and storage plants at Sukamandi and the selected seed districts.

TABLE 13.5 **Indonesia seeds project – Summary project cost estimates 1971/72 – 1975/76**

	Local	Foreign (Rupiahs million)	Total	Local	Foreign (US$ '000)	Total	Foreign exchange (per cent)
Ministry of Agriculture							
CRIA, Sukamandi Branch							
Buildings (houses, laboratories, offices)	57·4	24·6	82·0	151·8	65·1	216·9	30
Field and laboratory equipment	6·7	59·7	66·4	17·7	157·9	175·6	90
Vehicles	·8	4·8	5·6	2·1	12·7	14·8	86
Office equipment	·4	1·6	2·0	1·1	4·2	5·3	80
Irrigation infrastructure	39·7	19·2	58·9	105·0	50·8	155·8	33
Scholarships	–	18·5	18·5	–	48·9	48·9	100
Sub-total	105·0	128·4	233·4	277·7	339·6	617·3	55
Technical assistance	60·6	398·8	459·4	160·3	1 055·0	1 215·3	87
Counterpart staff costs[1]	210·3	–	210·3	556·4	–	556·4	–
Contingencies – physical	5·2	6·4	11·6	13·8	16·9	30·7	55
– price	10·8	10·5	21·3	28·6	27·8	56·4	49
Sub-total	391·9	544·1	936·0	1 036·8	1 439·3	2 476·1	58
Seeds control and certification service							
Vehicles	2·3	12·7	15·0	6·1	33·6	39·7	85
Buildings	20·8	4·0	24·8	55·0	10·6	65·6	16
Laboratory and office equipment	8·6	10·7	19·3	22·8	28·3	51·1	55
Contingencies – physical	1·1	1·8	2·9	2·9	4·8	7·7	62
– price	5·4	3·8	9·2	14·3	10·0	24·3	41
Sub-total	38·2	33·0	71·2	101·1	87·3	188·4	46
Total Ministry of Agriculture	430·1	577·1	1 007·2	1 137·9	1 526·6	2 664·5	57

National Seeds Corporation							
Irrigation infrastructure	438·0	211·2	649·2	1 158·7	558·7	1 717·4	33
Land development	25·5	28·8	54·3	67·5	76·2	143·7	53
Field machinery and equipment	61·8	349·0	410·8	163·5	923·3	1 086·8	85
Processing plants	127·1	719·0	846·1	336·2	1 902·1	2 238·3	85
Buildings, vehicles, office and workshop equipment	116·2	104·7	220·9	307·4	277·0	584·4	47
Sub-total	768·6	1 412·7	2 181·3	2 033·3	3 737·3	5 770·6	65
Technical assistance	111·6	666·5	778·1	295·2	1 763·1	2 058·3	87
Working capital[2]	100·0	–	100·0	264·6	–	264·6	–
Contingencies – physical	37·7	70·6	108·3	99·8	186·9	286·7	39
– price	136·7	209·4	346·1	361·6	554·0	915·6	61
Total NSC	1 154·6	2 359·2	3 513·8	3 054·5	6 241·3	9 295·8	67
Total project cost	1 584·7	2 936·3	4 521·0	4 192·4	7 767·9	11 960·3	65

[1] Expenditure comprises cost of technical assistance counterpart staff, training and staff overheads allocated to capital during five-year development period.
[2] Government contribution to cover initial NSC operating losses and to provide adequate working capital.

3 The mobilising of potential seed growers in the selected seed-growing districts.
4 The training of the required personnel to meet its staffing needs.
5 Marketing, distribution and trading in seed in general.

The staffing proposals for NSC comprised a President Director as the chief executive with supporting directors for production and processing, for technical services, administration and finance, and for marketing and sales.

At project preparation, the injection of foreign technical expertise to assist NSC in the operation was proposed. The problem was that according to the prevailing laws, the executive officers of a PERUM had to be Indonesian nationals. In other words, technical assistance experts could not enjoy executive authority. At the appraisal and credit negotiation stages, however, it was agreed that NSC should employ managing agents with executive responsibility for five years under the policy direction of the NSC Board.

The financing structure proposed was that NSC would derive its capital from the following sources: it would absorb the gross assets of the Sukamandi estate; the land and buildings sites for processing from the provincial authorities; other movable assets of the Ministry of Agriculture at Sukamandi; allocations from the government's annual development budget; and long-term loans from the government through the banking system at standard rates of interest. Additionally, government was expected to make a grant contribution to NSC's investment capital structure.

The use of consultants was proposed, to undertake the design work and supervise the execution of project works at Sukamandi to be carried out by contractors. The NSC would also appoint a consultant for final processing, plant design and supervision of installation. Contracts would be let for international competitive bidding for the supply and running in of equipment.

It was proposed that the Bank Rakjat (BRI) would undertake the credit-administration requirements of the project by obtaining from the Bank of Indonesia, credit facilities to permit it to extend a line of credit to the NSC to cover its trading requirements.

Project cost estimates and financing

The project cost estimates shown in Table 13.5 incorporate changes introduced during appraisal. Project costs totalled Rupiahs 4 521 million or US$12·0 million equivalent approximately, with a foreign exchange component of about US$7·76 million or roughly 65 per cent of total costs. Contingencies included in estimates consist of physical 5 per cent per annum, except for irrigation infrastructure for which 15 per cent was provided, and price contingencies of 5 per cent in the first year rising by 5 per cent each year to 25 per cent in year five to cover cost increases (local and import) in goods and labour.

It should be noted that the cost tables show planned expenditures by main categories, e.g. buildings, vehicles, irrigation infrastructure. Costs are above all shown under the responsible agencies, in this case, the Ministry of Agriculture, for the CRIA, Sukamandi Branch, and the Seeds

Control and Certification Service and the National Seeds Corporation. All items of cost are expressed both in local currency, and foreign currency (US dollars) and show the local and foreign exchange components.

The project was financed as follows: World Bank/IDA contribution of US$7·9 million of 66 per cent of total project costs, with the government providing the rest of US$4·1 million equivalent. Of the IDA credit, the NSC would receive US$1·7 million by way of technical assistance. Government's own contribution to NSC would amount to US$2·6 million equivalent also in equity and US$0·2 million equivalent towards technical assistance. Total NSC equity would thus be US$6·9 million.

The financing structure proposed for NSC – that is, that all funds be passed on as equity – arose because of the uncertainties involved in setting up an entirely new industry and because NSC would be able to service only a small proportion of its capital at the interest rate of 1 per cent per month, then prevailing and with repayment over 15–20 years. Any substantial interest payments would severely affect NSC's liquidity requirements. The financing requirements for the project by the Ministry of Agriculture would be met by IDA credit funds of US$1·5 million and government budgetary allocations of about US$1·2 million.

The evaluation phase

The financial analysis

The most significant elements are the impact on the seed producer and the NSC.

The seed producer's benefits would derive from two sources, namely increased yields and better prices for seed as against grain. Farmers growing a rice-seed crop in seed districts were expected to obtain 4 tonnes of dry paddy per ha as against their current yields of 3·5 tonnes per ha. These yield increases would come from better quality stock seed, minor irrigation improvements, higher levels of farm inputs and better supervision by extension workers. The crop of wet stalk-paddy would be

TABLE 13.6 Pro forma farm budget (Rupiahs)

Item	Without-project	With-project
Cost		
Seed[1]	1 000	2 000[1]
Roguing (4 times)	–	3 000
Pest control and insecticide	3 000	5 750
Other costs	42 000	42 000
Total	46 000	52 750
Gross income	68 680	85 680
Net income	22 680	32 930
US$ equivalent	60·0	87·0

[1] Increased seed costs arise from higher costs of stock seed purchases from NSC.

TABLE 13.7 Indonesia seeds project – National Seeds Corporation cash Flow (Rupiahs million)

	1971/72	1972/73	1973/74	1974/75	1975/76	1976/77	1977/78	1978/79	1979/80	1980/81	1981/82	1982/83	1983/84	1984/85	1985/86		Total
Income																	
Sale of certified seed	54·4	166·7	308·9	512·8	708·6	1 128·1	1 181·4	1 181·4	1 181·4	1 181·4	1 181·4	1 181·4	1 181·4	1 181·4	1 181·4		13 512·1
Sale of consumption rice	3·0	9·1	16·8	28·0	39·1	50·0	52·4	52·4	52·4	52·4	52·4	52·4	52·4	52·4	52·4		617·1
Receipts from rental farm	11·1	11·3	10·7	10·8	15·1	18·3	18·3	18·3	18·3	18·3	18·3	18·3	18·3	18·3	18·3		242·0
Total income	68·5	187·1	336·4	551·6	762·8	1 196·4	1 252·1	1 252·1	1 252·1	1 252·1	1 252·1	1 252·1	1 252·1	1 252·1	1 252·1		14 371·7
Expenses																	
Field operations	39·4	91·9	148·5	244·5	365·1	482·7	517·9	517·9	526·6	531·0	531·0	531·0	531·0	531·0	531·0		6 120·5
Processing and storage	23·1	53·6	84·5	155·0	210·8	272·7	287·8	287·8	305·8	315·6	315·6	315·6	315·6	315·6	315·6		3 574·7
Administrative overheads and non-seed activities	27·6	42·5	50·8	66·3	80·3	95·0	95·4	95·4	95·4	95·4	95·4	95·4	95·4	95·4	95·4		1 221·5
Stock seed from CRIA Sukamandi Branch[1]	15·0	4·5	10·4	14·8	19·4	66·5	66·5	66·5	66·5	66·5	66·5	66·5	66·5	66·5	66·5		729·1
Seed certification fee to seed control and certification service	·4	·8	1·1	1·5	1·9	1·9	1·9	1·9	1·9	1·9	1·9	1·9	1·9	1·9	1·9		24·7
Total expenses before taxes and depreciation	105·5	193·7	295·3	482·1	677·5	918·8	969·5	969·5	996·2	1 010·4	1 010·4	1 010·4	1 010·4	1 010·4	1 010·4		11 670·5
Income surplus (or deficit)	(37·0)	(6·6)	41·1	69·5	85·3	277·6	282·6	282·6	255·9	241·7	241·7	241·7	241·7	241·7	241·7		2 701·2
Corporation tax[2]	–	–	–	–	–	–	–	68·7	63·1	56·7	56·7	56·7	56·7	56·7	56·7		472·0
Income surplus (deficit) after taxes	(37·0)	(6·6)	41·1	69·5	85·3	277·6	282·6	213·9	192·8	185·0	185·0	185·0	185·0	185·0	185·0		2 229·2
Inflow																	
Surplus income	–	–	41·1	69·5	85·3	277·6	282·6	213·9	192·8	185·0	185·0	185·0	185·0	185·0	185·0		2 272·8
Government:																	
IDA credit 195–DND (irrigation)	107·1	–	–	–	–	–	–	–	–	–	–	–	–	–	–		107·1
Contribution in local currency	107·0	–	–	–	–	–	–	–	–	–	–	–	–	–	–		107·0
Proposed IDA credit – equity	401·3	164·0	516·9	412·4	119·8	–	–	–	–	–	–	–	–	–	–	1 614·4	2 588·7
Government's contribution to proposed credit – equity	253·6	187·0	243·8	169·9	52·9	51·9	15·2	–	–	–	–	–	–	–	–	974·3	
Total inflow	869 0	351·0	801·8	651·8	258·0	329·5	297·8	213·9	192·8	185·0	185·0	185·0	185·0	185·0	185·0		5 075·6

Outflow																
Income deficiency	37·0	6·6	–	–	–	–	–	–	–	–	–	–	–	–	–	43·6
Capital investment under proposed project[3]	554·9	351·0	760·7	582·3	172·7	51·9	15·2	–	–	–	–	–	–	–	–	2 488·7
Irrigation financed from credit IDA 195–DND	214·1	–	–	–	–	–	–	–	–	–	–	–	–	–	–	214·1
Replacements and renewals	–	–	–	–	–	165·2	165·2	165·2	165·2	165·2	165·2	165·2	165·2	165·2	165·2	1 652·0
Total outflow	806·0	357·6	760·7	582·3	172·7	217·1	180·4	165·2	165·2	165·2	165·2	165·2	165·2	165·2	165·2	4 398·4
Surplus (or deficit)	63·0	(6·6)	41·1	69·5	85·3	112·4	117·4	48·7	27·6	19·8	19·8	19·8	19·8	19·8	19·8	677·2
Cumulative surplus	63·0	56·4	97·5	167·0	252·3	364·7	482·1	530·8	558·4	578·2	598·0	617·8	637·6	657·4	677·2	–
Working capital requirements at 4 months' operating expenses	35·0	64·0	97·0	159·0	224·0	303·0	312·0	312·0	329·0	333·0	333·0	333·0	333·0	333·0	333·0	–

1 First year's purchase is from external source. CRIA, Sukamandi Branch will not supply until 1972/73. Purchase price 1972/73–1975/76 Rps 45/kg, from 1976/77 onwards Rps 137/kg.
2 45% on income surplus less depreciation with losses carried forward.
3 Total NSC less technical assistance, irrigation in 1971/72 and working capital Rps 100 million. Investment during the first five years: Rps 2 421·6.
4 Heavy costs arise from two cropping seasons in each year.
5 Working capital in year 1 is Rps 100 million less income deficiency of Rps 37·0 million.

purchased by NSC at a premium of 25 per cent over local prevailing prices.

A seed district farmer would have a net income from 1 ha of about Rupiahs 31 000 (US$87 equivalent) as compared with Rupiahs 23 000 (US$60 equivalent) for a farmer producing consumption rice as shown in the pro forma farm budget above. It was considered that such a potential increase in income was an adequate incentive to permit selected seed growers to enter and remain in the seed-growing business.

The budget, shown in Table 13.6, shows the situation before the project and with the project for a 1-ha farm yielding 4·32 tonnes wet paddy, producing consumption rice, and is compared with a similar farm producing rice seed as part of the project.

The project's financial impact on the NSC is shown in the cash flow for the enterprise in Table 13.7.

The projections of operating results show that the self-generated funds would be sufficient to cover only a very small loan. This provides, in fact, the justification for proposing a government equity contribution to NSC of US$6·9 million equivalent. While NSC would develop a small operating surplus of about Rupiahs 20·0 million from year eight onwards, the large working capital requirements for its trading activities would make it very unlikely for NSC to be in a position to make even modest dividend payments. The seed-pricing assumptions for sales of NSC seed were that project seed would be sold at Rupiahs 45 per kg wholesale and Rupiahs 55 per kg retail. This was modest compared to the prevailing rice-seed prices of about Rupiahs 36 per kg, while farmers were prepared to pay up to Rupiahs 40 per kg for seed which they considered to be good-quality seed.

Thus, while the project could not be described as being financially attractive in its early years, in that returns on invested capital were low, yet, the overall economic benefits were such as to warrant the country's undertaking the project.

The economic benefits

The project's main benefits would be the increased rice production brought about by higher paddy yields from the new seed stream with the consequent savings in foreign exchange on rice imports. The output of the new seed stream would, at full development in 1978, include about 258 tonnes of stock and foundation seed, and about 21 500 tonnes of cleaned and certified seed. Based on the assumptions about yield increases shown below, the incremental benefits were estimated at some 320 000 tonnes of rice annually, resulting in foreign exchange savings estimated at about US$28 million per year.

The project's economic rate of return, was estimated to be 58 per cent which is very favourable. The main assumptions were as follows:

1 The project would have a life of 15 years commencing in 1971/72 when first investments are made.
2 The price of imported coarse rice[1] would be US$87 per tonne c.i.f. Djakarta in the late 1970s.
3 All costs are net of import duties.

Benefits

These derive from the increased yields of farmers using the high-yielding varieties of rice seed of the new seed stream. They are estimated conservatively to be as shown in Table 13.8.

TABLE 13.8 Yield estimates (per cent)

	Yields with the HYV from the new seed stream	Yields with saved non-HYV
First crop after seeding	110	100
Second crop after seeding	107	100
Third crop after seeding	105	100

All seed produced would be sold in areas served by the BIMAS *nasional* and INMAS programmes where current yields are about 1·7 tonnes per ha. Other benefits derive from increased milling yields resulting from a more homogeneous grain (estimated to be 4 per cent); the better utilisation of farm inputs; seed requirements reduced from 40 kg/ha to 25 kg/ha; and additional research facilities at CRIA, Sukamandi Branch. These additional benefits have not been taken into account.

Costs

All project costs, both investment and recurrent, are included together with equipment and vehicles purchased in 1970 (but not financed by the project). An item has also been included to cover the cost of seed distribution (Rupiahs 10 per kg or US$26 per tonne) and for additional milling and handling charges (Rupiahs 7 per kg or US$19 per tonne).

Economic rate of return

Assuming these costs and benefits, the rate of return of the project is 58 per cent. With a drop in price to US$80 per tonne and US$70 per tonne the rate of return would be 53 per cent and 45 per cent respectively.

Sensitivity tests have been made on the following assumptions.

1 Seed production at Sukamandi only is successful and by year 1977/78 has reached 11 790 tonnes; seed production in the seed districts is found to be a failure after the full investment is made at Klaten; the consequent rate of return would be 47 per cent.

2 Only 50 per cent of Sukamandi production is sold as seed, 5 900 tonnes, the balance being disposed of as consumption grain; there is no production from the seed districts, and the full investment is made at Klaten; the rate of return would be 23 per cent.

Performance during implementation

Achievement and problems

During the implementation phase the project experienced serious difficulties, which resulted in considerable time and cost overruns and some seed quality problems. Actual performance, therefore, fell considerably short of targeted performance in respect of basic indicators of project output, such as the area under production, the area with irrigation infrastructure, and the area developed for mechanised agriculture at the Sukamandi estate – all factors contributing to a volume of production of certified seed far below expectations.

With respect to the seed districts, only one district had been developed at the end of the disbursement period, and both area of production and volume of certified seed production fell far short of planned targets. The main bottlenecks may be grouped under procurement, technical, managerial, financial and institutional reasons.

The procurement problems refer to delays by contractors in execution of project works. For example, land development was considerably slowed down by faulty topographical surveys, and delays in the construction of irrigation and infrastructure facilities. Procurement of equipment for land-development work suffered delays. The procurement and construction of seed-processing plant facilities also suffered undue delays for which cost escalation was only partly responsible. Temporary facilities for handling the crop at Sukamandi were strained to the limit. The recruitment of consultants to undertake planned breeding and research activities at Sukamandi also slipped considerably.

Technical problems had many origins. Probably the most significant, which expressed itself in the form of low yields (average of the high-yielding rice varieties was 2·5 tonnes/ha) resulted from the land-levelling operations involving earth movements. About 700 m^3/ha of earth was moved on the average – an operation which proved slow and costly. The movement of the top soil resulted in depressed soil-fertility levels despite high levels of fertilisation. In areas previously cropped with paddy, equipment tended to get bogged down, as mechanised cultivation proved difficult.

Lower percentage germination of seed was traced to two factors, namely, the slow rate of land development led to continuous cropping without a fallow as previously envisaged, and the changing of varieties on the same fields led to physical admixture. Furthermore, the temporary facilities for seed drying and storage could not cope with the production, and resulted in high-moisture contents, which retard germination. The continuous cropping also induced serious rat-infestation and control was proving difficult.

However, there were positive achievements on the technical side. Combine-harvesting proved quite successful, and as earlier envisaged, this had done much to relieve the labour shortage existing in the area at harvest time. Although seed quality, for the above reasons, was adversely affected, however, the quality was sufficiently superior to create a strong demand from distributors and farmers.

Managerial problems developed at both the senior and lower levels. There were delays in filling some top management posts with Indonesian nationals. Furthermore, NSC's staff was, in part, recruited from Ministry of Agriculture personnel, who lacked a commercial orientation and experience in the management of large-estate production. At the lower levels, inadequate workshop facilities and maintenance arrangements to service a large collection of machinery, while partly a financial problem also reflect a managerial deficiency.

Financial problems proved a major constraint. The NSC failed to get budgetary contributions from government at the levels requested. Meanwhile its operating and cumulative losses were substantial and made serious inroads into its total equity. Two factors principally accounted for this: the slow pace of development, resulting in considerably reduced levels of output, and the onset of rampant world inflation which resulted in very sizeable increases in project costs to levels unforeseen at the time of project preparation and appraisal. The NSC thus experienced problems both on the cost and the income side. The result of this financial squeeze is that the scale of the project had to be trimmed down, especially involving a reduction of the number of seed districts planned.

A further problem on the institutional side was the continuation of two parallel seed streams, contrary to what was envisaged. Thus agricultural extension stations and research stations continued to grow and sell seed, including sales to BIMAS farmers. The NSC had built up a good distribution network and dealers were able to dispose of all NSC-supplied seed to farmers, who were prepared to pay a premium for this over the seed from the government stations. Although not all the NSC seed was certified, the point is that farmers had grown to identify this as good or quality seed. Government however, was still faced with the problem of deciding which seed stream should be given priority.

A discussion

The basic question to be asked at this stage is whether the original project was sound in the light of later development. The fact that the project failed to achieve its planned rate of development, is not the most critical consideration in assessing the soundness of the project's concept. There were mitigating elements in the picture – for example, the oil price boom experienced by Indonesia in its exports meant that contractors were faced with new opportunities and were no longer interested in relatively small contracts. Furthermore, the budgetary constraints faced by NSC in its operations also reduced the possible rate of development.

The cardinal feature of the project's concept was that there was a strong demand for quality seed in Indonesia which could best be satisfied by a new seed stream involving rigid controls at all stages from research through production, processing, marketing and distribution. In retrospect, it is production rather than demand that has proved to be the bottleneck. The fact that all the NSC marketed production had been sold, and distributors, dealers and farmers were clamouring for more, at prices above other competing seed streams, suggests that the new stream could considerably expand its share of total demand for seed, if production could be stepped up considerably without sacrificing quality levels.

It will be recalled that the project's concept also involved the twin approaches of estate production and specialised seed-growing districts. The risks involved in the estate approach were early recognised. However, while problems have developed, the error was not in deciding in favour of a mechanised approach, but lay instead on the optimistic assumption about the rate at which benefits would accrue and the implications this would have for cost overruns. The depressing effects on yield that land levelling would involve and the time lag required for the soils to settle back to a more normal condition were not adequately anticipated. It is now clear that the project design considerably over-emphasised the role that estate production of seed should play in the process.

With regard to the seed-growing districts, it is further established that these were indispensable if Java's seed requirements were to be covered. For example, it is economically not feasible to service east Java requirements from Sukamandi because of the high transport costs.

It was not foreseen, further, that the NSC management would be so fully absorbed with Sukamandi that they would be unable to give sufficient attention to the development of seed districts. The inference from all the foregoing is that the scale of the project had been too ambitious, for a first-phase development, again not because of demand but because of production considerations.

Lastly, the strength of vested interests in preserving the old seed stream was under-estimated. The existence of two parallel seed streams, and the budgetary constraints of the NSC are elements of the same syndrome. In retrospect, the project should have progressively absorbed the provincial seed stations into the new system, instead of hoping that they would spontaneously cease production. In other words, project planning should have paid more attention as to how key requirements of a new seed stream could have been grafted on to the on-going seed stream.

References

1 A population planning worker in Central Java told the writer that her clientele believed that a large number of children was a gift from Allah. She explained that the concept of family embraced much more than the child's parents. In fact, if the parents could not take care of the child, the latter's upbringing would be the responsibility of the parent's closer or more distant relatives. Family concept is therefore much more embracing than prevails in Western industrialised countries.

2 Abbreviation for *Bimbingen Massal Swa Sembada Bahan Makanan* literally meaning: guidance for self-sufficiency in food stuffs.

3 Unpolished rice between 35 and 42 per cent brokens similar to Indonesian consumption rice.

14 Jamaica: agricultural credit

The case study, set in Jamaica, illustrates the steps involved in bringing an agricultural credit project up to the stage of appraisal by the financing institution. There are two aspects common to all credit projects: first, the lending institution's capacity to undertake the project must be appraised, and second, the composition and size of the lending programme must be developed and the terms of lending to final beneficiaries must be proposed.

The study begins with the identification phase during which the main emphasis is on achievements and structure of the country's agriculture, and on the analysis of performance of relevant institutions in the country. This diagnosis of the situation provides the basis for the decision-making phase, when it must be decided whether a case exists for proceeding to full preparation of the project(s). While two projects were identified in this case, one for owners of small farms, and one for owners of large farms, the case study takes the reader through the detailed preparation phase of the large-farmers' project only. The discussions embrace all the dimensions of a project, as discussed in Part III.

The identification phase

The task at identification was to review agricultural credit in Jamaica and to identify a credit project suitable for financing by an international financing agency.

The general setting

Jamaica's economy, primarily agricultural, has since the early 1950s undergone a marked transformation, largely because of the growth of the bauxite mining industry, the development of tourism, and the growth of light manufacturing encouraged by incentive packages designed to attract industry. These developments had rather profound effects on the agricultural sector. The more lucrative and prestigious job opportunities created thereby, introduced a greater disenchantment with agriculture among the agricultural population. This affected adversely the attitudes of potential new entrants into the agricultural work force of the country. A further consequence of this was the impact on values of agricultural land. The concessions granted to the bauxite mining companies, effectively meant that a substantial part of the land was taken out of the reach of the market. The purchase of land on Jamaica's north coast by predominantly North Americans for tourism development at extremely high values, considerably pushed upwards the value of agricultural land since in a small country such as Jamaica, about 11 400 km^2 (roughly 240 km long and 80 km at the widest point), it is impossible to isolate the effects of such important developments in one area from the rest of the country. The discriminatory tax incentive measures for industry *vis-à-vis* agriculture meant that investment funds, where possible, were channelled away from agriculture into industry.

The impact of all this was a lower growth rate in agriculture than in the rest of the economy. In fact agriculture remained relatively stagnant for several years. Agriculture's share of the GDP declined from 31·0 per cent in 1950, to 11·6 per cent in 1965, and to 9·0 per cent in 1969. Yet the sector is important in terms of employment, engaging 40 per cent or the largest share of the labour force. With the high unemployment and under-employment then prevailing in Jamaica, developments in the agricultural sector can have strong political and social implications.

Trade in agricultural products worsened. Food imports increased noticeably, rising from US$50 million in 1964 to US$72 million in 1969. The rising imports, largely meat and fish products, reflect not only increases in income levels, but also derive from larger inflows of tourists, with their distinct preference for imported foodstuffs.

Some features of Jamaican agriculture

Jamaica's agriculture is characterised by dualism, with a well-organised estate sector lying side by side with a peasant subsistence sector. The former is modernised, employing modern technology with purchased inputs. The sugar cane industry, for example, has earned a well-established reputation for the quality of its research work on sugar cane production. Yields for estate-grown crops, e.g. sugar cane and bananas, were markedly higher than in the peasant sector, with hardly any technological spillover occurring from the former to the latter.

While Jamaica's agricultural projection is fairly well diversified, as seen in Table 14.1, sugar cane production still predominates. Land-use figures show that of the total land area of 1·1 million ha, about 0.22 million ha are cultivated, while grassland occupies about 0·20 million and secondary scrub or ruinate accounts for about 0·16 million ha.

Sugar cane dominates the cropping pattern with about 73 300 ha,

TABLE 14.1 Jamaica – Agricultural output and value of exports: 1967–69

	Production			Value of exports (J$ '000)		
	1967	1968	1969	1967	1968	1969
Sugar ('000 tonnes)	449	445	383	20 176	34 004	28 586
Bananas ('000 tonnes)[1]	192	153	151	13 332	13 794	12 469
Citrus ('000 boxes)[2]	1 233	1 445	1 219	5 336	5 060	6 716
Copra (tonnes)	20 395	18 475	17 217	–	–	–
Cocoa (tonnes)	1 676	2 344	1 491	544	1 030	1 049
Pimento (tonnes)	1 905	2 246	3 225	2 666	3 030	4 241
Coffee ('000 boxes)	1 449	932	1 130	486	746	676
Beef (tonnes)	–	15 281	14 126	–	–	–
Pork (tonnes)	–	5 388	6 050	–	–	–

[1] Includes exports only and no domestic consumption.
[2] Only for processing and excluding fresh fruit consumption.

with production located on the best soils and in the most accessible areas. The industry had its labour problems, with labour shortages and unemployment existing side by side. The physically demanding work, especially at harvest time, is very unattractive to the farm-labour work-force, especially at prevailing wage rates. There is further a 'snob' value involved in such work and workers would seek employnent in remote areas so that their friends would not know that they had been cutting cane. The sugar industry then included 18 sugar estates, while some 20 000 smallholders accounted for just over 50 per cent of the cane production. Government at the time was then considering a more liberal attitude towards mechanisation of the industry because of profitability problems and the artificial labour shortage.

Banana exports had also experienced declines. Bananas, like sugar cane, are also an estate and a smallholder crop. They are grown either as pure stands or in conjunction with other crops, especially coconuts. Some 65 000 growers are involved with about 32 800 ha. The industry is particularly important because of the weekly income package it generates. Coconuts account for about 38 450 ha. The coconut plantations on the north coast were being severely devastated by a disease called lethal yellowing disease. Other important crops include cocoa, 17 400 ha; pimentos, 11 740 ha; citrus, 10 900 ha; and coffee, 8 500 ha.

The cattle population stood at some 400 000 head, with about 200 000 ha classified as grazing areas. Jamaica had pursued a cattle-breeding scheme and it pasture development work was well recognised.

Farm size

The farm-size distribution according to the government's 1968 census, is shown in Table 14.2.

TABLE 14.2 Jamaica: Farm-size distribution

Farm size (ha)	**Number of farms**	**Hectarage of farms**	**Hectarage per farm**
0–2	144 604	90 577	0·6
2–9	36 881	134 982	3·6
10–39	3 000	50 628	16·8
40–199	699	60 091	84·7
200 and over	295	273 720	928·4

Some 185 000 farms had a total hectarage of about 0·6 million ha. The highly skewed land-ownership distribution reflected in the above figures is typical of Latin American countries. Thus, 95 per cent of the total farms fell in the category of less than 10 ha but accounted for only 37 per cent of farmland. About 145 000 farms were less than 2 ha with an average size of 0·6 ha. It is clear, therefore, that the large bulk of the farming population depended on economically non-viable farming units. This has important policy implications which are discussed later.

Government programmes and policies

Between the period covering the middle 1950s through to the late 1960s, the government's approach to increasing agricultural production involved both loans and grants or subsidies to farmers, totalling about US$32·1 million equivalent[1], of which about US$22·3 million was for loans with the remainder as subsidies. The three major schemes, namely, the Farm Development Scheme (1955–60), the Agricultural Development Programme (1960–62), and the Farmers' Production Programme (1963–68), each had loan and subsidy elements.

An assessment of performance showed that as of early 1969 these schemes were running into financial difficulties, as loans outstanding stood at US$15·1 million of which about an estimated 40 per cent was in arrears. The problem was thus examined in relation to the structure of production. It is important to note that the smallholder sector is predominantly responsible for domestic production of food crops with the estate sector concentrating on export crops. Jamaica's growing food demand was not being met and the rising food import trend was a source of concern. While the government recognised that the schemes had a large social element, through their subsidy component, it was concluded that these schemes had failed to generate the expected increases in production. The government concluded that a major defect of these schemes was the intermingling of loans and subsidies. The objectives of social support through subsidies and increased agricultural production would thus have to be treated differently.

It was decided, therefore, that farms in the range of less than 2 ha could not operate as self-supporting viable units, and could not qualify for credit on commercial terms. Social considerations dictated, however, that they should continue to receive subsidies, and the capital estimates for 1969/70 provided for about US$0·72 million for this purpose.

Recognising that farms ranging from 2 to 10 ha had the potential for increased food production, the government developed a policy for credit and intense extension support to boost production on such farms and to make them financially viable. This group would, therefore, not qualify for subsidies. Resulting from this policy decision, the Parliament approved in May 1969 the following: a subsidy scheme; a new self-supporting loan assistance scheme, and the establishment of regional land authorities to undertake the required development activities. The Ministry of Rural Land Development was assigned responsibility for those tasks. The self-supporting loan assistance scheme envisaged that farms in the 2–10 ha group could generate net annual incomes of about US$1 000. The government budgeted about US$1·0 million for this credit programme to be conducted in about five pilot areas, involving about 400 farmers in the first year.

The major thrust would be on the development of existing farms, which could qualify for credits of about US$2 700 per farm; new farmers were not excluded but their higher development costs would qualify them for credits of US$6 700 per farm. Concessionary lending rates of 6 per cent were proposed with an amortisation period of 20 to 25 years. The scheme targeted for some 10 000 farms being involved over five years and envisaged external financing for the proposal. The small-farmers project, later discussed, was based around this scheme. Farmers owning more than 40 ha were excluded from participation in this scheme.

The institutional setting

The plethora of institutions involved in agriculture does pose problems for any comprehensive type of development project which falls within the domain of different agencies. The problem becomes one of coordination, and the most logical way to achieve this is to ensure that those most directly concerned are brought in at the project-formulation and project-implementation phases.

Ministry of Agriculture and Fisheries

The continuous whittling down of the functions of this Ministry had resulted in a situation whereby the only remaining functions it exercised in practice were agricultural and livestock research, training, irrigation development (very modest in the Jamaican context), fisheries and some agricultural planning and dissemination of information. The various agricultural stations carried out plant propagation and distribution activities in addition to research.

The ministry's major weakness was that it did not and could not exercise the role of spokesman for the government on important policy issues in Jamaica's agriculture. While it is true that the Minister of Agriculture chaired the various commodity boards, the fact that each Industry Board had carved out its sphere of influence, meant that Ministry officials at the working level could only marginally influence fields of activity lying outside of their day-to-day operations. The role of key officials was largely one of making a technical input on an *ad hoc* basis into the problems of a particular industry. Their contribution at the level of policy issues was minimal or non-existent.

Possibly, the function that suffered the most, within the context of Jamaica's agricultural administrative system, was extension. The staff of the various Boards, e.g. citrus, cocoa, coffee, coconut, etc., were so involved in the administration of schemes, distribution of credits, or grants, inputs, etc., that technical know-how and technology transfer – an extension function – were relegated to the background. Furthermore, the recent development, whereby the extension function of the Ministry of Agriculture was handed over to the new Ministry of Rural Land Development, reinforced the predominant role that agriculture would play in research. Unfortunately, however, no mechanism had been developed or was contemplated whereby the Ministry's research findings would reach the farm level through the multiplicity of extension networks prevailing. The Ministry's main departments comprised agronomy, agricultural chemistry, plant protection, livestock research and development, veterinary services and agricultural planning.

Ministry of Rural Land Development

Established in February 1969, the Ministry was charged with several functions, previously carried out by the Ministry of Agriculture. The new agency would spearhead agricultural development on smallholdings principally. The following departments were transferred from the Ministry of Agriculture to the new ministry: extension, forests, lands, survey, and two commissions, namely the Land Development and Utilisation Commission and the Watershed Protection Commission.

The new Ministry (MRLD) would be responsible for administering farmers' development programmes, including the small farmers loan assistance scheme. It would also administer the programme for the creation of some 11 new Land Authorities, which would cover most of Jamaica's smallholder farming population. While some coordination with the Ministry of Agriculture at the headquarters level was planned for, since the new Ministry would also be served by the Agricultural Planning Unit of the Ministry of Agriculture, what remained unclear was the mechanism for coordination of activities at lower levels, especially the farm level.

The Land Development and Utilisation Commission

Set up under the Land Development and Utilisation Act of June 1968, the Commission was established because a significant part of the cultivable and potentially productive agricultural land in Jamaica was either uncultivated or used at levels below its production potential. Accordingly, the Act sought to ensure that cultivable lands above a certain area (initially 40 ha), then idle or under-utilised, should be put into cultivation or an acceptable form of utilisation, and to provide the process by which the Ministry of Agriculture might acquire such lands if the owners did not undertake the needed improvements according to an acceptable development plan.

The Commission's work provided a starting point for the later elaboration of the large-farmers' project. Land-owners with land adjudged idle had to submit development proposals and by mid-1969 about 240 development plans had been submitted out of 320 land-owners served 'notices'. Where owners refused to take steps to develop their properties, the Commission was empowered to 'certify' this to the Minister of Agriculture who was then authorised to acquire the land. This step of land acquisition had only been used in two cases. The Commission's work, however, led to sales of property to new owners, and some subdivision of properties.

The Agricultural Development Corporation

Established in 1952, the Corporation undertook the following developmental activities: large-scale demonstrations and trials with new crops; commercial production of crops in short supply; dairy development including imports of stock; the running of a farm mechanisation scheme. For various reasons the Corporation suffered a loss in operations, especially in its machinery services. However, this is an interesting case where a government employs the 'corporation' approach to undertake innovative testing, development and promotion of production activities and agricultural services, deemed to be in the public interest.

The land authorities

They were corporate bodies established under the Land Authorities Law of 1954. Their *raison d'être* was to administer the more depressed agricultural areas. Their sphere of influence was thus limited to specific geographic areas. Their main task was to ensure the efficient utilisation of land within the areas, including reforestation and afforestation and other soil-conservation activities. To this end, the authorites were empowered to administer government loan and subsidy schemes.

Organisationally, the authorities were administered by a board with an executive chairman and fell under the portfolio of the Minister of Rural Development after transfer from the Ministry of Agriculture. Their sources of funds were varied and comprised: funds from general revenues for administration, contributions from the Development Fund; loan funds from the Agricultural Credit Board, and other contributions from the Coffee Board. By 1969, they had expended funds, including loans and subsidies, totalling about US$7·2 million.

Their activities included provision of infrastructure, roads and air-strips for crop spraying; installation of village water supplies; soil conservation; pest control; afforestation; provision of farm housing units; and the operation of handicraft and home economics centres. They not only operated the extension services, but ran special projects to provide farmers with improved stock, both animal and crops; nursery supplies, etc. They acquired land for subdivision and land-settlement purposes. They engaged in marketing activities and operated the food-crop purchasing centres set up by the Agricultural Marketing Corporation. They also ran cool storage facilities for potatoes. Their very substantial credit and subsidy operations are discussed under the section of credit institutions.

The assessment of performance of the two authorities, then operating, was limited by the lack of reliable data. It was estimated that one authority had tripled the value of production between the five-year periods of 1950–55 and 1960–65. Apart from obvious production increases, there were visible but unquantifiable project benefits from the erosion-control measures. The most convincing element was visual: it was the contrast between the farmers and farms in the project areas and those outside or contiguous to the project areas. The government of Jamaica, itself satisfied with the achievements, saw the authority approach or regional development areas as the answer to the development of the non-estate sector of Jamaica's agriculture. The total of thirteen regional land authorities to be created, would execute this task. It is in this context that proposals for a credit project specifically aimed at small farmers were analysed.

The credit institutions

The more important ones include the Bank of Jamaica, the commercial banks, the Agricultural Credit Board, the land authorities, the Jamaica Development Bank.

The Bank of Jamaica is the country's Central Bank. This fully government-owned bank in its role of monetary authority, issues currency, controls interest rates through setting of the bank rate (then 6 per cent) and controls the liquidity of the banking system by imposing a minimum ratio of liquid assets/deposits on the commercial banks, in addition to exercising control over certain types of credit. If offered re-discount facilities for first-class papers (then at 7 per cent) for an initial nine-month period with the possibility of a similar extension.

The commercial banks provided an important but fluctuating source of loan funds to agriculture, primarily to the more affluent and larger farmers. As of the end of 1968, the seven banks had outstanding loans totalling about US$209 million of which about US$9·6 million was for agriculture, with a further US$14·4 million for agricultural processing industries. Two banks did most of the agricultural lending with loans for sugar cane dominating (about 42 per cent) and the remainder for cattle, bananas and other crops and pigs and poultry. However, it is quite possible that these figures for agriculture were understated since many borrowers with established credit-worthiness did not indicate the specific end-use of their borrowings or diverted them to other purposes including developing their farms.

Generally, the lending terms were for a maximum of three years, while there was a trend by one institution to longer-term loans specially for processing activities. The collateral requirements varied according to the borrower's stature. Interest rates also varied, ranging from 0·5 to 2·0 per cent above the prime rate. The prevailing rates thus ranged between 7 and 9 per cent, in addition to a 1 per cent commitment fee.

The Agricultural Credit Board was government's main institution for small and medium-scale lending to agriculture. Its formation in 1960 was designed to bring some sense of order and control over the cooperative rural loan societies (also called People's Cooperative Banks) existing since the turn of the century.

Its main tasks were to promote agriculture by supervising, controlling and making loans and grants to the 115 cooperative banks; disbursing loans to the nine approved organisations, including the land authorities, etc.; and making direct loans to farmers.

Its resources derived from various sources, e.g. government funds for loans and relief purposes, including hurricane rehabilitation; the government's agricultural credit-revolving fund; and foreign borrowings, e.g. USAID loans.

The Board's operations were managed by a board of directors appointed by the Minister of Agriculture, holding tenure at his pleasure. The Board's current expenditures budget was an integral part of the Ministry of Agriculture budget. The Board's staff held full civil service status. While the Board was supposed to enjoy some semi-autonomous status, in reality the Minister exerted such strict control over the Board that it lacked any independent stature in its operations. A manager was the Board's chief executive officer.

The Board's volume of loans outstanding stood at about US$15·1 million for the fiscal year 1968/69, of which about 80 per cent was accounted for by loans to the People's Cooperative Banks, 15 per cent to approved organisations and 5 per cent by its own direct loans to farmers. The People's Cooperative Banks were in turn separate legal entities and autonomous credit societies whose members subscribed share capital with liability to the extent of their contributed shares. They had their own management committee and their own staff. They borrowed from the ACB at 3 per cent and on-lent to their members at 6 per cent. Their loan repayment performance was extremely unsatisfactory. Between 1965 and 1969, the percentage of total loans in arrears increased from 15 to 40 per cent. In fact, an 'operations review' of the cooperation banks, concluded that 77 of the 115 People's Cooperative Banks were bankrupt and past the stage of possible survival. Among reasons for failure were lack of local leadership, poor management, loans to farmers with poor credit standing, and to farm units that could not possibly be viable.

A reorganisation of these banks was thus in process, aimed at consolidating the more successful and better located ones to a more manageable 30–35 units. This would enable better supervision by the ACB which by 1969 had directly taken over the management of 30 People's Cooperative Banks.

Land authorities These also undertook credit operations. For example, the Christiana Area Land Authority had disbursed about US$1·06 million between 1956 and 1960, of which total loan repayment at the end of the period averaged just slightly over 50 per cent, meanwhile for the Yallahs Valley Land Authority, with its total loans of US$0·37 million between 1951 and 1969, the repayment was 60 per cent. Since the ACB was merely the channel through which loan funds to the authorities were funnelled, the authorities disbursed and collected their loans, and returned to the ACB the loan repayments and interest. Collateral for loans could take the form of first mortgage on land, or two personal guarantors; crop liens or bills of sale. Loans included short-term up to two years, and medium-term up to seven years.

The loan-recovery performance was far from satisfactory. For the Christiana Authority it was estimated that as of March 1969, 69 per cent of the loans outstanding were overdue. In fact, defaults in loan and interest payments appeared to have been permanent features of the credit operations; the authorities themselves in their annual reports had stressed the need to intensify collection efforts. Legal action was finally taken against some 17 loan defaulters but recovery was minimal.

The Jamaica Development Bank (JDB)

This is more fully discussed under the large-farmer's project. Formerly, the Development Finance Corporation, it was reconstituted as the Jamaica Development Bank, whose major lending fields would comprise industry, tourism, housing and agriculture.

The decision-making phase

Identification established that there was a case for a credit project in Jamaica. However, the analysis of the situation suggested that there was merit in having two separate sub-projects, one aimed at small farmers and the other at large farmers. The conclusions were based on the following.

Owners of large farms In this case the virtual absence of long-term institutional credit was a major constraint to agricultural development. Generally, this class of farmers depended on rolling over of annual overdraft facilities to finance their operations both capital and current.

Obviously such short-term financing is not conducive to the successful implementation of a long-term development plan especially one involving long-gestating plantation crops. Because the opportunity to tap commercial bank funds was largely a function of the prevailing liquidity of the commercial banks at any one moment – and this fluctuated widely – and because other economic sectors enjoyed priority in the banks' loans portfolios, agriculture was likely to suffer most in contractionary phases of the business cycle.

The role of credit in their case would merely be to serve as a stimulant to a wider and faster adoption of technology known to them. This refers in particular to the coconut and banana industries, where in both cases replanting programmes with new varieties, implied a temporary reduction in current production income levels, coupled with large capital outlays.

This group of farmers did not have an enviable situation, as they lay virtually outside the institutional source of funds administered by the Agricultural Credit Board on the one hand, and they exercised only a residual claim on the lending of commercial banks. The group referred to here excludes the largest enterprises with parent companies based abroad with wide access to capital.

A further major consideration was the imperative need for positive measures to overcome Jamaica's stagnation in agriculture. Rapid increases in production could come from this large-farmer sector, and a strong effective demand existed for loan funds. A large-farmer credit project would also give a psychological boost to this ailing sector, which felt strongly that government had not treated them as equals with the other economic sectors. It was expected that one impact of the new capital flows deriving directly from the proposed large-farmers credit project, would be to induce these farmers to increase their own investment in agriculture.

The question of who would administer it and how, had to be resolved. Government had proposed that the Jamaica Development Bank would be the appropriate agency. The suitability and capacity of the JDB to execute the project is examined later.

Owners of small farms The justification for a credit project aimed at small-farm owners and the possible format and scale of such a project were more difficult to formulate. In the first instance, the virtual absence of reliable data on the performance, as measured by production increases, of the various government programmes aimed at the small farmer over about two decades, meant that performance could not correctly be evaluated.

As opposed to the case of large farmers, where the removal of the credit constraints was expected to generate immediate production benefits, the small-farmer sector was plagued with so many problems, that the role of credit *per se* would be ineffective in the absence of other complementary measures. For example, a prominent feature of the ownership of small farms in Jamaica is the disorganised marketing of produce (with the possible exception of bananas). Certainly, marketing of food crops exhibited a whole host of deficiencies. The question of supervision to ensure that loan funds were in fact applied to the purposes for which they were intended, was a matter of concern. A strong teaching (extension) component would have to be built in, to teach farmers simple, improved practices, aimed at raising productivity, e.g. the application of purchased inputs. Furthermore, because hillside farming was largely a small-farmer operation, with the large estates occupying the flat and more gentle-sloping areas, planning had to take account of anti-soil erosion practices, e.g. contouring, and terracing, with strong implications for land development costs, especially for new enterprises or undeveloped parts of existing enterprises.

Having regard to all these constraints, the administrative costs of any scheme aimed at the small farmer would inevitably be higher than that for large farmers, as considerable supervision would be entailed.

Leaving aside for the moment the question of suitable organisational arrangements to implement a small-farmers' credit project, there were very positive elements which far outweighed the negative aspects above, and dictated that a case could be made for a small-farmers' project. The single most important argument was the government's own recognition that the intermingling of subsidies and credit had clouded the issues. Henceforth, the beneficiaries of direct subsidy schemes would be separated clearly from beneficiaries of credit schemes.

Credit programmes would, henceforth, concentrate on raising production and would have to stand the tests of financial viability of the proposed farm operations. This did not imply that government would forego the use of subsidy on an across-the-board, industry-wide or sector-wide basis if specific situations at any particular moment warranted it.

Preliminary farm models had demonstrated that such farms in the range of 2–40 ha could generate satisfactory financial rates of return, and that cash-flow patterns were adequate to cover loan amortisation and interest payments and leave the farmer substantially better off than he was before. There was evidence that investment in small farms could be economically justified.

The successes achieved by the land authorities clearly had shown the potential for increased production even under the most depressed conditions. However, any first-phase project should concentrate on the more favourable areas, with a good growth potential based on the available land, water and physical infrastructure resources. This should simplify the task of organisation, and should yield earlier demonstrable results to convince potential investors to undertake a larger project later.

Institutionally, the creation of the Ministry of Rural Land Development, including in its mandate extension services to small farmers, was clear evidence of the government's intention to tackle the problems of small farms in a concerted fashion. Budgetary and human resources had been allocated, and the support of an international financing agency could do much to help orient and channel government's efforts at resolving problems, in the right direction.

The small-farmers' project

The Agricultural Planning Unit (APU), which serviced both the Ministry of Agriculture and Forests and the Ministry of Rural Land Development, had carried out some studies on possible project areas that fell within the domain of the proposed new land authorities. These surveys provided sufficient basic data about the main agricultural features of the project areas, e.g. cropping patterns, land utilisation – cropped versus uncropped land – and farmers' development intentions so as to be able to propose the main outlines of a project, requiring further detailed preparation for financing by the Inter-American Development Bank. The main elements of the project are listed below.

1 The implementation over two years of agricultural development plans for five pilot areas, one in each of five regional land authority areas out of the 13 proposed authorities.
2 The development of 900 farms (primarily in the range of 2–10 ha and not exceeding 40 ha) through the provision of credit facilities and technical assistance. Credits would be applied for the provision of on-farm development, for farm buildings and water supplies.
3 Trained extension workers would provide intensive guidance.
4 The financing of agricultural machinery purchased by farmer-contractors with a guarantee that they would service their farmer neighbours under acceptable terms and conditions.
5 The provision of technical assistance to assist in executing the pilot project and developing plans for the expanded development in future phases of the project.

Project funds would be applied largely to the development of existing enterprises where the APU surveys had established that roughly 50 per cent of the areas of the farms surveyed were idle or partly idle. For each of the proposed sub-project areas proposed, model development plans were drawn up to reflect differences between areas in their crop and livestock husbandry practices; the ecology and production potential of the various soil classes; the market prospects of various alternative cropping possibilities; and the development wishes of the farmers as expressed by the farm surveys.

Very briefly, the organisational arrangements proposed for implementing this sub-project were: the project would be executed by the Ministry of Rural Development through the boards of the land authorities for the five proposed areas. The Agricultural Credit Board would be responsible for the banking aspects of the project. Its functions would be limited to the banking aspects of the project, which would include the approval or rejection of loan applications based on considerations of credit-worthiness, security and collateral, and repayment capacity as shown by the cash-flow projections. It would have prime responsibility for loan collection and station a credit officer in each of the authorities' areas.

The preparation phase – large-farmers' project

We examine in this section the three major components of any credit project: namely the institution, the lending programme and the terms and conditions of lending. Capacity of the institution is an all-embracing word covering such elements as management capability and staffing; financial resources and operating policies; historical performance in lending, etc.; and the extent to which the institution reinforces broader government policies of the overall development strategy.

The lending programme spells out the various lending fields – that is, what the loan funds would be applied for – and establishes loan targets by category. The composition of the lending programme is arrived at by examining several elements. The first element is to determine whether a technical basis exists to support the lending programme. Market and demand prospects for various crop and livestock commodities which are important in the country's agriculture and which have priorities in government's plans, are the second element. The third element is the attempt to establish the probable level of effective demand by lending category, that is the potential number of loan applicants who are able, and willing to borrow, under the proposed terms of lending. The possible number of loans and probable average loan size would thus give the total effective loan demand. The fourth element is financial and requires that the repayment capacity of borrowers be established by farm-model and cash-flow projections, demonstrating that there is sufficient incentive (i.e. incremental gains deriving to the farmer from the project) to induce the farmer to borrow. The findings here have to be tailored to what the lending institution can realistically be expected to achieve during a stated time period, having regard to any support that the institution may reasonably expect from other agencies with an interest in the project. That is, the question of administrative limitations.

Finally, the terms and conditions of lending refer to such elements as the interest rate; the length of the repayment period including grace periods; the existence or not of a commitment fee on approved loans; the question of who bears the foreign exchange risk; the requirements of collateral and security; limitations on lending to individuals, or corporate entities, and the supervision and reporting requirements of the institution from its clients.

The Jamaica Development Bank (JDB)

Since the Bank was a recently established institution, it did not have prior experience in the field of lending to agriculture. Founded in March 1969, under the Jamaica Development Bank Act, it became the successor to the Development Finance Corporation, a fully government-owned institution established in 1959 to provide funds for developing the tourism, industry and housing industries. The JDB took over the Corporation's resources. The changes introduced at the time of the take-over included: an increase of authorised capital, by US\$7·2 million to US\$12 million, with the government subscribing US\$2·4 million and the Central Bank US\$4·8 million; the government take-over of several large loans in arrears; the opening of a new window for lending to agriculture and top management changes. As of December 1969, the issued and paid-up capital was US\$5·4 million.

Under the Act establishing the JDB, it was empowered to borrow, with the limitation that its total liabilities must not exceed seven times its paid-up capital and reserves. It was authorised to borrow through issue of debentures or bonds. As at end 1969, its borrowings stood at US$10·8 million, with a debt/equity ratio of 1·5:1. Its foreign borrowings, negotiated but not disbursed, stood at US$10 million approximately.

Its loans as at end 1969 totalled US$15·6 million of which US$5·8 million was to wholly owned subsidiaries. It held equity in two companies, totalling US$1·8 million. Financial projections, made of JDB's portfolio and profits, suggested a healthy profit position after, say, five years, and that no debt-servicing problems were foreseen.

The JDB was managed by a board, nominated by the Minister of Finance, and comprising 11 members including the chairman, general manager, and representatives from the Central Bank and Ministry of Finance. The private sector was also represented on the board of directors. The general manager, previously with the Ministry of Agriculture, had a very experienced foreign adviser in credit banking to assist.

The Bank's staffing comprised 19 professionals, well-qualified but with little banking experience. The project staff comprised four senior officers, one each for industry, tourism and agriculture and a general economist. The officer leading the Agriculture Credit Section was an experienced agriculturist, well versed in tropical agriculture, both in Jamaica and other Caribbean territories.

The JDB had issued a statement of policy which constituted firm guidelines as to how it would operate. These covered scope of operations, e.g. industrial enterprises, tourist enterprises, and agricultural enterprises operating on a commercial basis. Bank assistance would take the form of loans, equity investments, guarantees, underwriting of security issues, and issuing of debentures, and preference and ordinary shares.

The minimum size of loans by the JDB would be US$12 000 equivalent except in agricultural loans where the minimum would be US$24 000 equivalent. The Bank would seek to obtain adequate collateral, as security on its loans, while being satisfied with the viability of the project and the earnings potential of the enterprise. General lending principles covered such items as the financial viability of projects, repayment periods and repayment capacity, and working capital.

It is obvious that, in this particular case, the institution had not the time to establish its own performance record. In such a situation, it becomes critical to assess thoroughly the lending programme. The objective of the large-farmers' project was, therefore, not only to obtain production increases but to build up the JDB as an institution so that it could, in the future, undertake with confidence a larger workload.

The lending programme

At preparation, the projects' total costs, including contingencies and a technical assistance component, were estimated at J$7·04 million (US$8·45 million). On the assumption that loan beneficiaries would contribute a minimum of 20 per cent of development costs, then JDB's actual lending would total about J$5·5 million (about US$6.6 million).

The lending by category would be as follows:

TABLE 14.3 Jamaica: Lending programme by year and by category (J$ '000)

Category	Year 1	Year 2	Year 3	Total
Beef	420	630	840	1 890
Dairying	164	164	164	492
Citrus	363	363	363	1 089
Coconuts	259	259	388	906
Bananas				
Farms	94	188	188	470
Nursery	63	62	–	125
Others	100	200	200	500
Total	1 463	1 866	2 143	5 472

The main operations that would be financed under the project would be the expansion of operations of 45 beef farms with required pasture establishment and farm buildings; the expansion of 15 dairy farms; establishing 1 214 ha new citrus orchards; establishing 1 416 ha of Malayan dwarf coconuts; replanting 1 012 ha of existing bananas to Valery bananas and setting up 80 ha of banana nurseries, and establishing limited amounts of other crops.

The initial screening process

Data from three sources were employed in arriving at decisions about what categories should be included in the lending programme. First, the Land Development and Utilisation Commission (LDUC) had developed land use data on about 1 300 large farms totalling 327 530 ha. Since the beginning of its activities in 1966, the following crop development programmes had, by 1969, taken place on some 597 farms:

TABLE 14.4 Jamaica: Land use data on large farms

	Hectares
Sugar cane	865
Bananas	711
Citrus	823
Coconuts	2 995
Improved pastures	5 392
Food crops	1 453
Others	1 210
Total	13 449

These findings in fact reflected the larger farmers' own assessment of the market prospects and profitability of various development possibilities. Two items, namely sugar cane and food crops, not included in the lending programme, are discussed below.

The Agricultural Planning Unit of the Ministry of Agriculture had undertaken a survey of the possible interest of large farmers (i.e. farms exceeding 40 ha) to borrow from the JDB. Of the survey sample of 500, replies were received from just over 200 farmers. Half of these latter reported no interest in loan funds either because they could do their own financing or had access to alternative sources, e.g. commercial banks. Their expressed credit needs are shown in Table 14.5.

TABLE 14.5 Jamaica: Sample survey findings of credit requirements

Category (J$)	Number of farmers	Credit requirements (J$)
Under 10 000	29	199 450
10 000–20 000	20	239 000
20 000 and over	48	3 078 300

Thus, loans requests in the largest category averaged about J$60 000 (US$72 000) per loan. The main requests are shown in Table 14.6.

TABLE 14.6 Jamaica: Loan requests by crop category

	Hectares
Pastures	2 578
Mixed crops	1 469
Sugar cane	9 340
Citrus	8 535
Bananas	6 904
Coconuts	4 862
Coffee	243

While the data also provided a useful guideline, the limitations of the responses must be recognised, since the respondents did not then know the proposed terms of lending that would be offered by JDB.

The third source of data was information gathered in interviews with industry bodies and associations. The Citrus Growers Association, for example, canvassed its leading members in the main citrus-growing areas of Jamaica. The objective was to establish 'effective demand' – that is, a realistic appraisal of the potential number of borrowers, and amounts they would be willing to borrow.

The data from all available sources were used to put together a list of criteria for developing the lending programme.

1 The product should be relatively important in the national economy.
2 The agronomy for the particular crop should be well established, or a new technology exists.
3 The product should have satisfactory demand and market aspects and should not be subject to unduly wide price fluctuations.
4 Farmers should have demonstrated a strong interest in expansion of the commodity.
5 The availability of funds from other sources, including commercial banks, for particular commodities should be a factor.

The lending programme proposed, thus excluded four items: sugar cane, cocoa, coffee and food crops, based on the above considerations. Omission of sugar cane in particular, was sharply questioned by government and Central Bank officials, since it was Jamaica's most important crop. However, the ailments of the sugar industry were so profound that an injection of credit – even though it were substantial – would not correct the industry's basic structural problems, also, prevailing market uncertainties made it inopportune to consider heavy capital injections at this stage. Two emerging developments were taken into account – namely the Commonwealth Sugar Agreement would expire, and that it would be impossible, then to forecast the impact of Britain's possible entrance into the European Common Market or Britain's future relationship with her Commonwealth and West Indian sugar suppliers. The industry's high cost structure was leading to the closing down of several small sugar factories; many estates were reducing their sugar cane hectarage, and labour for harvest was becoming increasingly difficult. With regard to financing requirements, the larger estates were owned by expatriates who ment their own capital requirements. Domestically, the commercial banks had given by far the largest part of their agriculture portfolio to the sugar industry. It could hardly be argued that a shortage of funds was the most significant constraint facing the industry.

While it must be pointed out that the international financing agencies were at the time quite wary about financing the sugar industry largely because of the depressed prices prevailing, it is important to note that credit, in and by itself, could not be expected to achieve miracles in the Jamaican context, The industry's problems required first a thorough economic study with the objectives of spelling out measures for rationalising the industry, and setting it on a more profitable basis through cost reduction and productivity-increasing measures, including the rational consolidation of the processing establishments.

Cocoa was omitted because it suffered from marked production declines resulting from depressed prices which made it far less attractive to growers than other alternative crops. Furthermore, it was largely a peasant crop confined to one parish. Coffee faced a very uncertain demand. Food crops were omitted from this sub-project, largely because they are in the small-farmer domain; and received considerable attention in the small-farmers' project.

The coconut sub-project

This sub-project is used as a model to demonstrate the steps in the process of detailed preparation of the lending programme. We discuss below, the industry's situation and structure, on-going programmes including the technical basis, marketing and demand, the sub-project proposal, and the financial aspects.

General situation

While coconuts are grown throughout Jamaica, they are heavily concentrated in high-rainfall zones of the eastern parishes. There were an estimated 4·8 million bearing coconut trees (about 38 000 ha). Production is highly concentrated, with the larger estates producing the bulk of the nuts for copra. For example, about 2·4 per cent of all growers (215 out of 8 725) delivered two-thirds of the total crop for processing. Marked differences existed in the yield levels of larger versus small producers, the former averaging about 30 per cent higher than the latter.

The main variety grown is the Jamaican tall, usually with a plant population of about 98 trees per ha, with yields of about 30 nuts per tree per year. Intercropping with bananas is common practice, while other crops, citrus or pimento, are also used. Other varieties grown in considerably less quantities are the Malayan dwarf or the 'San Blas' or 'Panama'. While other oil seed crops are grown in very small quantities, largely experimental, coconuts constitute the main source for vegetable oil production.

About 84 per cent of coconut production feeds into processing, with the rest going for fresh domestic consumption. Between the period 1954 to 1967, copra production had grown at roughly 3·5 per cent per year, with the 1967 production being about 18 200 tonnes. The drought years 1966, 1967, 1968 had taken a toll on production, a situation further accentuated by the effects of lethal yellowing disease.

Jamaica's north coast, especially the parishes of St Mary and Portland, once offered an unbeatable spectacle of scenic beauty dominated by coconuts. The sight of dead and dying palms, as a result of the above disease, is a threat not only to the landscape but to Jamaica's coconut industry. The disease, long latent in Jamaica, was known to produce isolated deaths in the west of the country. About 1961, it first appeared on the north-east coast, and began its unchecked spread throughout the plantations. The industry estimated that losses were then running at about 100 000 bearing trees annually. On-going programmes to control this are discussed below.

Industry structure and on-going programmes

The Coconut Industry Control Law of 1945 established two organs namely a Coconut Control Authority with a general supervisory role over the industry, and a Coconut Industry Board (CIB) to regulate marketing, research and programmes affecting the industry. The Board membership comprised nine members, four government and five private.

The Board's major functions comprise: exclusive trading rights in copra, including running its own subsidiary processing companies, and sale of copra to other processors, and control over imports and exports of processed coconut products. It provides technical assistance to growers, undertakes research, administers a windstorm insurance scheme and the lethal yellowing disease replanting programme. It supplies planting material to farmers, and distributes fertilisers at cost to its registered growers. The Board's overall performance was convincing enough to ensure that it could play a vital role in ensuring the success of the sub-project.

Research programme This was well documented and showed a high level of professional competence, and their data provided the technical basis for the formulation of the sub-project. The research programme, operating since 1961, had two major objectives: to develop improved methods for coconut production including use of fertilisers, pesticides, intercropping, etc., and selection and breeding programmes aimed both at yield and disease resistance to lethal yellowing disease. Their work had demonstrated clearly that the Malayan dwarf coconuts, while not entirely immune to lethal yellowing disease, possessed extremely high resistance even in interplantings with the susceptible Jamaican talls. The few isolated cases of infection and mortality were attributable to hybrids cross-pollinated with pollen from the talls. Their work with new planting methods was an important element in project preparation – that is, the replanting of the new dwarfs in pure stands. The existing system of interplanting with bananas gave yields of about 0·98 tonnes of copra per ha per year, while the new planting method resulted in doubling of yields and much larger net income per ha.

The Research Department was strongly staffed, including three expatriate scientists, 30 field assistants and six laboratory assistants and the Department received further assistance from FAO and the Ministry of Overseas Development of the UK.

The replanting programme operated by the Board, since 1968, was based on Malayan dwarf coconuts. The dwarf is less prone to hurricane damage, bears earlier and yields about 17 300 nuts per ha as compared to the Jamaica talls with 3 000 nuts per ha; dwarf nuts are smaller, however. The 1968 planting totalled 243 000 plants, while 345 000 plants were targeted for 1969. The five-year target 1968–73 was set at 1·75 million new palms, or a total area equivalent to about 3 035 ha, about a half of which would include coconut groves devastated by the disease. The Board provided assistance to registered growers, equivalent to roughly a quarter of the replanting costs. The Board's funds for implementing this programme were rapidly drying up.

A windstorm insurance scheme was operated by the Board to protect growers against loss of producing trees from hurricanes and floods. Benefits received on the average by growers amounted to about 50 per cent of losses incurred.

Operating results The Board's operating results were quite satisfactory. In 1966 and 1967, it showed trading profits of J$42 000 and J$35 000 respectively, though it incurred deficits in 1968 and 1969 because of the lower throughput, largely drought-induced. Its resources derive from the difference between its purchase price to growers and its sales of copra to processors which was about J$21.80 per tonne of copra.

Demand and prices

The Board estimated that the demand for copra and oils and fats was growing at roughly 5–6 per cent per year, while copra production was averaging a growth rate of 3·5 per cent over the past decade. The Board further estimated that the current gap was about 15–20 per cent short of domestic demand requirements or roughly 5 000 tonne of copra. Imports of copra and oil (expressed as tonnes, copra equivalent) were about 1 500 tonnes each in 1964 and 1965 while the 1969 import figure had reached 5 000 tonnes to cope with the growing deficits.

Projections for 1980 were that copra demand requirements would rise from the prevailing base of 19 000 tonnes to about 33 000 tonnes. If these demand levels were to be met, a new planting programme would have to reinforce the replanting programme.

Prices are fixed by law and remain in force over a seven-year period. The latest prices introduced in 1967 were as follows: J$20.40, per 1 000 nuts of Malayan dwarf, and J$25.00 per 1 000 nuts of Jamaican talls. These are grower prices for husked nuts delivered at any copra drier. The copra purchasing prices was J$152.75 per tonne while the Board sold at J$174.55 per tonne, giving it a margin of J$21.80 per tonne for its various operations and administration costs.

The final copra price was distributed as follows: grower's share was 72·4 per cent; copra processor 17·6 per cent and the CIB share 10 per cent. Of the latter 2·3 per cent went to windstorm insurance fund, 3·1 per cent to research; 0·6 per cent lethal yellowing replanting fund, and 4 per cent to the Board's operations.

TABLE 14.7 Jamaica: Investment and operating costs for coconuts interplanted with bananas

Investment items – Year 1	**Cost per ha (J$)**
Land preparation	98·9
Single ripping and harrowing	46·9
Soil conservation or drainage	29·6
Planting material	98·9
Fertilisers	48·0
Pesticides	13·5
Roads	49·5
Farm equipment	64·2
Buildings	24·7
Labour costs	152·2
Contingencies, including supervision	60·5
Total	686·9
Operating costs – Year 2	
Weeding	35·8
Pruning	9·9
Drains upkeep	24·7
Fertiliser application	14·2
Harvesting	29·6
Fertilisers	64·2
Pesticides	13·5
Roads upkeep	12·3
Contingencies, including supervision	19·8
Total	224·0
Years 3 and 4	91 per annum
Years 5 and 6	76 per annum
Year 7	55·5 per annum
Subsequent years	50·5 per annum

The sub-project proposal

On the basis of an assured demand, firm guaranteed prices, an efficient Industry Board with demonstrated performance, and a variety fully tested for cultural requirements and disease resistance, there was a strong case for including coconuts in the lending programme. The sub-project proposed, therefore, aimed to establish 1 416 ha of Malayan dwarfs, a figure later increased at the appraisal stage.

The sub-project would support the Board's on-going planting programme, which would have to be scaled down considerably in the absence of contributions from the government. About 80 per cent of project funds would be devoted to replanting programmes in areas ravaged by, or vulnerable to, attack by lethal yellowing disease in the main coconut-producing belt, while the other 20 per cent would be in areas less prone to hurricane damage, that is the eastern parishes.

It was estimated that there should be a minimum of 35 requests for loans, over a three-year period, with the majority of the requests averaging well over 40 ha each. Surveys of planting intentions carried out during preparation showed, for example, that four such large growers would establish over 800 ha over three years if loan funds were available to supplement their own resources. It will be recalled that the data had shown a stratified production in which the large bulk of copra originated from the larger farmers. Farm models were, therefore, based on 40 ha, which were quite realistic.

Project farms would be located in areas deemed suitable by the Jamaican Development Bank after appropriate consultation with the Board's technical staff. Project funds would be used to promote the establishment of pure stands with a plant population of 250–300 plants per ha, in contrast to the existing mixed stands of 100 talls per ha. Loan funds would cover up to 80 per cent of project costs, and would be used for land preparation, planting materials, construction of farm roads, machinery and equipment, purchased inputs, e.g. fertilisers and pesticides, and labour. Interplanting of bananas would be a project cost, so as to recover some establishment costs but bananas would be progressively taken away as the coconuts reached maturity. Supplies of selected planting material were not a constraint to obtaining the modest requirements of the proposed planting programme.

Financial aspects

The establishment costs of a unit – 40 ha, on which the farm models were based – totalled J$32 350 and included the costs for the first two years. Operating costs, i.e. maintenance costs until the coconuts reached full maturity, were not capitalised as it could be shown that they could be largely offset by the returns from bananas intercropped. The development of the 35 such units would imply a total cost of J$1 333 million (rounded) or about US$1·6 million and with the assumption of 80 per cent loan, the JDB loans for coconuts would total J$906 000 (US$1·08 million).

The investment costs for year one, that is, the year of planting, are shown in Table 14.7 as J$686·9 per ha. The operating costs for later years are shown in the same table.

The planting and production schedules are shown in Table 14.8. Thus, on a model 40-ha farm, 20 ha would be planted in the first year and 20 ha in the second year. The same table shows yield assumptions. These were estimated to reach 6 000 nuts per tree by the seventh year, under the recommended package of cultivation practices.

Table 14.9 shows the gross value of production from a 40-ha enterprise as it approaches full maturity. As the banana intercrop is phased out, gross income would stabilise at J$12 000 by the eighth year.

The cash flow and financial rate of return calculations are shown in Table 14.10.

TABLE 14.8 Jamaica: Coconuts with bananas interplanted – Planting and production schedules

Planting schedule	**Year 1**	**Year 2**	
Bananas 1st 20 ha	20		
2nd 20 ha		20	
Coconuts 1st 20 ha	20		
2nd 20 ha		20	
Production estimates	**Year 1**	**Year 2**	**Year 3**
Bananas 'Lacatan' (kg/ha)			
Full stand	4 480	10 090	13 450
Reduced stand	3 380	7 480	10 125
	Year 5	**Year 6**	**Year 7**
Dwarf Coconuts (units per ha)	6 175	12 350	14 825

Interest rates were calculated at two levels, namely 8·5 per cent and 9·5 per cent to assess the repayment capacity of borrowers under varying interest assumptions. The data show that this sub-project would generate an internal rate of return of 23·5 per cent.

The other sub-projects

Similar exercises were undertaken in respect of the beef, dairying, citrus and banana enterprises to derive a three-year lending programme.

A special word is merited about the banana sub-project. The industry had some 65 000 growers and its 1968 exports of J$13·8 million (US16·56 million) represented about 22 per cent of agricultural exports. Apart from marketing problems, the industry had a high cost structure due to low overall productivity. Internal transport costs were high and the Banana Board which was experiencing severe financial difficulties, had to collect fruit from everywhere.

Basic structural changes were necessary if the industry were to become economic. For example, zoning of production in the more favoured areas with proximity to major shipping points, would considerably limit the areas to be provided with services, such as spraying, and thus reduce costs. However, the industry had at its command a tool to raise productivity very rapidly at the farm level, namely a high-yielding, wind-resistant variety, the 'Valery', whose attributes constituted a sound technical basis for launching a 'Valery' replanting programme.

In consideration of this, it was decided at the project-preparation stage to introduce a sub-project with the modest objective of spearheading a change-over from existing banana varieties to 'Valery' in selected zones. Planting material and the appropriate input packages for establishment of 1 000 ha were proposed at a total cost of J$795 000 (US$0·954 million). Both the Ministry of Agriculture and the Banana Board supported this proposal. However, it was recognised at the time of

TABLE 14.9 Jamaica: Production estimates per farm (40 ha)

	Year 1	Year 2	Year 3	Year 4	Year 5	Year 6	Year 7	Year 8
Bananas (kg × 1 000) 1st 20 ha	69	154	205	205	143	162	–	–
2nd 20 ha	–	69	154	205	205	143	62	–
Total	69	223	359	410	348	305	62	–
Gross value (J$)	2 525	8 100	13 125	15 000	12 750	7 500	2 250	–
Coconuts (nuts × 1 000) 1st 20 ha	–	–	–	–	125	250	300	300
1st 20 ha	–	–	–	–	–	125	250	300
Total	–	–	–	–	125	375	550	600
Gross value (J$2 per 100)	–	–	–	–	2 500	7 500	11 000	12 000
Total gross value (J$)	2 525	8 100	13 125	15 000	15 250	15 000	13 250	12 000

TABLE 14.10 Jamaica: Finance projections (40-ha model)[1] (J$)

Item	Year 1	Year 2	Year 3	Year 4	Year 5	Year 6	Year 7	Year 8	Year 9	Year 10	Year 11	Year 12	Year 13	Year 14	Years 15–35
Cash inflow															
Total sales	2 525	8 100	13 125	15 000	15 250	15 000	13 250	12 000	12 000	12 000	12 000	12 000	12 000	12 000	12 000
Loan funds	12 160	13 720	–	–	–	–	–	–	–	–	–	–	–	–	–
Farmers contribution	3 040	3 430	–	–	–	–	–	–	–	–	–	–	–	–	–
Total cash inflow	17 725	25 250	13 125	15 000	15 250	15 000	13 250	12 000	12 000	12 000	12 000	12 000	12 000	12 000	12 000
Cash outflow															
Operation costs	3 400	7 950	9 100	9 100	8 350	7 725	6 775	5 290	5 050	5 050	5 050	5 050	5 050	5 050	5 050
Investment costs	11 800	9 200	–	–	–	–	–	–	–	–	2 600	–	–	–	–
Total cash outflow	15 200	17 150	9 100	9 100	8 350	7 725	6 775	5 290	5 050	5 050	7 650	5 050	5 050	5 050	5 050
Cash balance before debt service (total cash inflow – total cash outflow)	2 525	8 100	4 025	5 900	6 900	7 275	6 475	6 710	6 950	6 950	4 350	6 950	6 950	6 950	6 950
Loan amortisation	–	–	–	–	2 588	2 588	2 588	2 588	2 588	2 588	2 588	2 588	2 588	2 588	–
Outstanding loan	12 160	25 880	25 880	25 880	23 292	20 704	18 116	15 528	12 940	10 352	7 764	5 176	2 588	–	–
Interest payment: 1st alternative 8·5%	1 034	2 200	2 200	2 200	2 200	1 980	1 760	1 540	1 320	1 100	880	660	440	220	–
Interest payment: 2nd alternative 9·5%	1 155	2 459	2 459	2 459	2 459	2 213	1 967	1 721	1 475	1 229	983	738	492	246	–
Cash balance after debt service: 1st alternative	1 456	5 990	1 825	3 700	2 112	2 707	2 127	2 582	3 042	3 262	882	3 702	3 922	4 142	6 950
Cash balance after debt service: 2nd alternative	1 330	5 641	1 566	3 441	1 853	2 474	1 920	2 401	2 887	3 133	779	3 624	3 870	4 116	6 950
Accumulated Cash balance: 1st alternative	1 456	7 446	9 271	12 971	15 083	17 790	19 917	22 499	25 541	28 803	29 685	33 387	37 309	41 451	48 401
Accumulated Cash balance: 2nd alternative	1 330	6 971	8 537	11 978	13 831	16 305	18 225	20 626	23 513	26 646	27 425	31 049	34 919	39 035	45 985
Financial rate of return															
Total sales	2 525	8 100	13 125	15 000	15 250	15 000	13 250	12 000	12 000	12 000	12 000	12 000	12 000	12 000	12 000
Operating costs	3 400	7 950	9 100	9 100	8 350	7 725	6 775	5 290	5 050	5 050	5 050	5 050	5 050	5 050	5 050
Investment costs	11 800	9 200	–	–	–	–	–	–	–	–	–	–	–	2 600	–
Total costs (operating + investment costs)	15 200	17 150	9 100	9 100	8 350	7 725	6 775	5 290	5 050	7 650	5 050	5 050	5 050	5 050	5 050
Balance for calculating the interest rate of return (total sales – total[3] costs)	(12 675)	(9 050)	4 025	4 025	6 900	7 272	6 475	6 710	6 950	6 950	4 350	6 950	6 950	6 950	6 950

[1] Bananas used as catch crop.
[2] Replacement of farm equipment J$2 600 every 10 years.
[3] The internal rate of return is about 23·5 per cent over 35 years.

preparation that the implementation of this sub-project would be conditional on a satisfactory resolution of the marketing crisis then prevailing. This crisis developed when the major marketeer of Jamaica's fruit in the UK complained of losses incurred from poor fruit quality and low throughput levels resulting in under-utilisation of their shipping and marketing facilities.

Swift changes in government policy are not at all uncommon during the process of preparing an agricultural project. In the five-month intervening period between the preparation mission and the World Bank appraisal mission, the government introduced a very substantial programme of grants to the banana industry aimed at quickly stepping up the levels of production. Under such circumstances, the demand for loan funds would almost certainly have been negatively affected. Because of this, the appraisal mission correctly decided to omit the banana sub-project from the lending programme.

Terms and conditions of lending

At the time of project preparation, the JDB's lending rate for agriculture was not known, since it was just about to begin operations. Accordingly two alternative interest rates were used, namely 8·5 per cent and 9·5 per cent. The rate actually decided on later was 8·75 per cent per annum. Additionally, borrowers would have to pay a commitment fee of 0·5 per cent on the undisbursed loan balance, plus a service fee of 0·5 per cent before the first loan disbursement is made. These rates compare favourably with the 9·5 per cent then being charged to non-agricultural borrowers.

Jamaican farmers showed deep concern about levels of interest rate. Other lending conditions, e.g. repayment and grace periods, hardly appeared to provide sufficient incentive to overcome the overriding concern with the interest rate.

The problem of the on-lending rate, however, should be viewed not only in terms of its possible impact on farmer loan demand but also on the lending institution. In this case, one objective of the credit project was the building up of the JDB as an institution. This implies, among others, that it should be financially strong. The spread or margin between the cost of funds to JDB from the World Bank and other sources, and the rate at which it on-lends should permit a build-up of profits.

The question of who would bear the foreign exchange risk was discussed with the authorities. JDB officials maintained that the government should bear the risk. It was later agreed that the government would carry the foreign exchange risk but that it would receive from JDB a fee of one-tenth of 1 per cent on loans outstanding to sub-borrowers, such fee to be included in the 8·75 per cent rate to sub-borrowers.

On the basis of cash-flow tables, the grace periods and repayment periods shown in Table 14.11 were proposed.

It should be noted that citrus farms were suggested for the longest grace period. This reflects the fact that traditionally the Jamaican citrus grower does not intercrop his groves and so there can be no cash inflow until the groves attain bearing age.

TABLE 14.11 Jamaica: Grace and repayment periods by category

Category	Grace period	Years of repayment	Total
Beef	5	10	15
Dairy	4	8	12
Citrus	7	8	15
Coconuts	4	10	14
Bananas	3	5	8

The scale of the lending programme

The total project cost including technical assistance and contingencies totalled J$7·04 million (US$8·45 million) with an estimated foreign exchange component of US$3·6 million. The loan negotiated later for this project was US$3·7 million.

The main aggregates are shown in Table 14.12 for ease of discussion.

TABLE 14.12 Jamaica: Aggregates of the lending programme

	Unit project cost (J$)	Number of units	Total cost (J$ '000)	Total loan (J$ '000)
Beef farms	52 500	45	2 362	1 890
Dairy farms	41 500	15	615	492
Citrus orchards	45 400	30[1]	1 362	1 089
Coconut orchards	32 350	35[1]	1 133	906
Banana farms	26 750	25[1]	670	470
Banana nursery	–	–	125	125
Other crops	–	–	625	500
Total	198 500	150	6 892	5 472
Add: technical assistance			150	150
			7 042	5 622

[1] Each unit represents 40 ha.

Some senior government officials raised the question of the small size of loan that might result from such a submission. However, the proposed scale of the project was arrived at after intensive efforts to establish effective demand for loan funds, by category, by Ministry of Agriculture officials, by the various commodity boards and farmer associations, by commercial banks, and by JDB officials.

There was the problem of administrative constraints. The preparation and processing of an average of 40 loans in the first year, 50 in the second and 60 in the third year, would in fact be operating at the upper level of JDB's capacity. Credit demand on the one hand and the JDB's delivery system on the other hand had to be reconciled.

In any event, the JDB, with no experience in agricultural lending required time to build up not only staff but also to develop experience in this field before it could competently handle a lending programme of any large magnitude.

Organisation and management aspects

The JDB confirmed that only sub-borrowers with technically and economically viable farm-development plans would qualify for its financing. Apart from this requirement, the JDB would assess the ability of the applicant or his authorised manager to execute the approved development plan, and also the credit-worthiness of the applicant.

The head of the Agricultural Division would report to the general manager and loan approval would be a normal Board function. Professional staff would rise from four in the first year to eight by the third year. The disciplines would comprise expertise in livestock, agronomy, farm management, and agricultural economics. The objective was that JDB's own staff should have full responsibility for the preparation, appraisal and supervision of sub-projects.

The locally recruited group would be supported by two internationally recruited experts, to be financed under the World Bank loan, comprising a project economist with considerable experience in the economics and financing of agricultural projects, and a livestock expert experienced in animal production in the tropics.

The JDB undertook to enlist the support of staff and utilise the experience of other established institutions in the implementation of its agricultural lending programme. The Ministry of Agriculture, the Coconut Industry Board and the Citrus Board gave assurances that their staff would be available to assist JDB's staff in the technical evaluation of loan requests and in their preparation. They would also assist in supplies of planting material for project beneficiaries.

In order to ensure coordination between JDB's policies and loan operations and that of the involved institutions, e.g. ministries, government departments, industry boards and associations, the Agricultural Credit Board, etc., the government undertook to establish a coordination committee for this purpose.

Project's financial and economic benefits

The overall justification of the credit project was based on the following expectations.

1 An increase of gross farm value of production of crop and livestock products amounting to J\$2·64 million (about US\$3·17 million) annually at full development.

2 Foreign exchange savings estimated at US\$3·0 million annually, at full development.

3 A satisfactory rate of return to the economy of 15 per cent.

In addition to these returns at the national level, there would be farm-level benefits expressed as producer returns, and institutional benefits to the JDB.

The estimated production increases are shown in Table 14.13. While national production increases estimated were substantial, the project's success is, first and foremost, determined by expected net income increases at the level of the producers. In all cases, the major assumptions referred to yields, costs and prices. The results are shown in Table 14.14.

The rates of return were shown to be quite satisfactory and should induce farmers to undertake the proposed investments. Farmers would

TABLE 14.13 Jamaica: Production increases resulting from project

Item	Project production	Gross value (J$ '000)	Existing production	Percentage increase
Beef farms				
Carcass beef (kg × 1 000)	502	476	11 000	4·9[1]
Breeding heifers (head)	1 665	300	20 000	8·3
Dairy farms				
Milk ('000 litres)	2 537	223	39 777	6·4
Carcass beef (kg)	48 580	30	–	–
Breeding heifers (head)	29	5	–	–
Citrus farms				
Oranges ('000 boxes)	540	551	1 800	30·0
Grapefruit ('000 boxes)	600	300	700	85·7
Coconuts (million nuts)	21	420	135	15·6
Bananas (tonnes)	8 900	332	190 000[2]	4·7

[1] Includes carcass beef output of dairy farms.
[2] Represents 1967 exports as 1968 exports of 150 000 tonnes resulted from a bad drought.

TABLE 14.14 Jamaica: Farm-level benefits by lending category

Sub-project	Before development (J$)	At full development (J$)	First year full development	Estimated financial return on incremental investments (per cent)
Beef farms	2 390	13 100	7	18
Dairy farms	1 320	14 020	5	27
Citrus farms	–	20 975	11	17
Coconuts	–	6 710	8	23
Bananas	11 620	21 200	5	15

be able to service their loans adequately, and to meet income tax payments.

The project's economic rate of return was estimated at 15 per cent as shown in the Table 14.15. The sub-projects were assumed to have a composite economic life of about 20 years, with bananas at 12 years, beef and dairy enterprises at 20 years, and citrus and coconuts up to 35 years. Project costs include technical assistance and JDB's administrative costs. No shadow pricing was used for foreign exchange, or labour. Predicted world market prices for copra were used, and a subsidy on fertilisers taken into account.

Institutional benefits claimed were that the project would help to launch JDB into the agricultural lending field. The project would also help to remove a severe constraint to Jamaica's agriculture by providing a source of long-term credit, under acceptable terms and conditions, for sustained agricultural development of the larger producing farm units. The project would further bring into full production some 9 300 ha of land, then idle or largely under-utilised. Finally, the project would aim to introduce the most up-to-date technology of beneficiaries' farms, and hopefully would exert a 'demonstration effect' on adjacent farmers not participating in the project.

The appraisal process

During appraisal, the project concept remained substantially unchanged. Some changes in the composition of the lending programme were introduced, but, however, the total project cost was estimated at J$6·60 million, compared to J$6·89 million of the preparation mission.

The introduced changes consisted of: the elimination with government's agreement of the banana sub-project for which about J$0·6 million had been planned. The beef and dairy components remained practically the same. However, the coconut sub-project was substantially increased to cover 2 000 ha instead of the 1 416 ha proposed at preparation. Finally, a component for other crops totalling J$0·62 million proposed at preparation was excluded so as to simplify the lending programme.

At the appraisal stage, there were no significant project-related issues that could not then be resolved and that required settlement by the two parties – the World Bank and the government – at the time of loan

TABLE 14.15 Jamaica: Economic rate of return (J$ '000)

Year	Benefits (increased sales)	Investments	Increased operational costs	Administration	Total costs	Balance*
1	(8)	3 115	136	77	3 328	(3 336)
2	307	1 697	444	82	2 223	(1 916)
3	1 147	626	770	87	1 483	(336)
4	1 687	137	839	52	1 028	659
5	1 825	67	891	52	1 010	815
6	2 024	–	1 033	62	1 095	929
7	2 226	–	1 024	57	1 081	1 145
8	2 333	–	926	52	978	1 355
9	2 473	–	993	52	1 045	1 428
10	2 625	–	1 074	52	1 126	1 499
11	2 701	–	1 328	52	1 380	1 321
12	2 701	–	1 007	62	1 069	1 632
13	2 369	–	922	57	979	1 390
14–15	2 369	–	862	52	914	1 455
16	2 369	–	1 000	52	1 052	1 317
17–20	2 369	–	862	52	914	1 455

* The economic rate of return is 15 per cent over 20 years.

negotiations. In this case, the borrower was the Jamaican Development Bank with the government of Jamaica being the guarantor for the loan.

Performance during implementation

Viewed in terms of the rate of disbursements, the project had been quite successful, after an initial slow build-up period during which the institution had to get to know its clientele and vice-versa. For example, the sub-borrowers had to become familiar with the collateral requirements of the JDB. Meanwhile JDB had to develop its own working practices and procedures, e.g. for loan appraisal and disbursement.

The more important developments during the project's implementation were the shifts in demand for loan funds between lending categories in contrast to what was envisaged at the preparation and appraisal stages. The loan demand for beef and dairy farms was considerably larger than the loan funds allocated to these categories. This arose principally because of changes in government price policies for beef and milk, which, while admittedly low at the time of project preparation were not constraints to financial viability of such enterprises. In 1973 for example, the government lifted price controls on domestic beef cuts, resulting in much higher beef prices in general, with a marked incentive effect on cattle production, and reflecting itself in increased local demand.

Meanwhile, loan demand for coconuts fell considerably below expectation, because subsidies from the Coconut Industry Board were available to coconut growers for replacing Jamaica talls with the Malayan dwarfs. As a result of these shifts in demand, the JDB, after obtaining the World Bank's agreement, effected shifts of loan funds between the different categories.

Finally, one new development, not envisaged at the time of preparation, was the proportion of new farmers, who would have figured among JDB's clientele. In fact the lending programme was based on the idea that major emphasis would be given to the expansion of operations on existing enterprises. In sharp contrast, about half of the loans were to new farmers developing uncultivated land, and existing farms.

The first-phase project had led to several 'repeater projects'. Favourable government policies have contributed to this success. The project could have benefited initially from a built-in control system to monitor results at the farm level.

References

1 The prevailing exchange rate was US\$1·00 = J\$0·80, or J\$1·00 = US\$1·20.

Bibliography

W. Allan, *The African Husbandman*, Oliver and Boyd, Edinburgh, 1965.
W. C. Baum, The Project Cycle, in *Finance and Development* (a publication of the International Monetary Fund and the World Bank Group), No. 2, 1970.
T. F. Carroll, *Issues of Financing Agrarian Reform: The Latin American Experience*, World Land Reform Conference, FAO, Rome, WLR/66/5, 20 June–2 July 1966, pp. 18–19.
S. J. Chand, *et al.*, *Technical Change in Asian Agriculture*, Australian National U.P., Canberra, 1973.
P. Dasgupta, S. A. Marglin and A. L. Sen, *Guidelines for Project Evaluation*, New York, United Nations/UNIDO.
J. C. de Wilde, *et al.*, *Agricultural Development in Tropical Africa*, Vols I and II, Johns Hopkins University Press, Baltimore, 1967.
René Dumont, 'The Biggest Famine in History Has Just Begun', in *Hunger Politics and Markets, the real issues in the Food Crisis*, ed. Sartaj Aziz, New York University Press, New York, 1975.
FAO, *General Guidelines to the Analysis of Agricultural Production Projects – Agricultural Planning Study*, No. 14, 1971.
FAO, *Agricultural Development in Nigeria, 1965–1980*, Rome, 1966.
FAO/World Bank Cooperative Programme, *The Muda Study*, Vols I and II, FAO, Rome, 1975.
J. P. Gittinger, *Economic Analysis of Agricultural Projects*, The Economic Development Institute, International Bank for Reconstruction and Development, Johns Hopkins University Press, Baltimore, 1972.
J. P. Gittinger (ed.), *Agricultural Projects, Case Studies and Work Exercises*, Economic Development Institute, International Bank for Reconstruction and Development, Washington, D.C., 1973.
J. R. Hansen, *A Guide to the UNIDO Guidelines: Social Benefit/Cost Analysis in Developing Countries*, UNIDO, 31 July 1975, ID–75–6769 (unpublished).
A. O. Hirschman, *Development Projects Observed*, The Brookings Institute, Washington, D.C., 1967.
L. Joy and P. Payne, *Food and Nutrition Planning*, Consultants' Report Series No. 35, FAO. ESN:CRS/75/34, Rome 1975, p. 63.
J. A. King (Jr), *Economic Development Projects and their Appraisal*, The Economic Development Institute, Johns Hopkins Press, Baltimore, 1966.
R. Krishna, Agricultural Price Policy and Economic Development, in *Agricultural Development and Economic Growth* (eds.), H. M. Southworth and Bruce F. Johnston, Cornell University Press, Ithaca, New York, 1969, pp. 497–540.
H. Koontz and C. O'Donnell, *Principles of Management*, McGraw Hill Book Co. Inc., 1959.
U. Lele, *The Design of Rural Development. Lessons from Africa*, a World Bank Research Publication, Johns Hopkins University Press, Baltimore and London, 1975.
W. A. Lewis, *Development Planning*, New York, 1966.
I. Little and J. A. Mirrless, *Social Cost/Benefit Analysis*. Vol. II of *Manual of Industrial Project Analysis in Developing Countries*, Development Centre Studies, Paris: Development Centre of the Organization for Economic

Development and Cooperation, 1969.
S. A. Marei, *The World Food Crisis*, Longman, London and New York, 1976.
J. W. Mellor, *The Economics of Agricultural Development*, Cornell University Press, Ithaca, New York, 1966.
J. W. Mellor, *The New Economics of Growth*, A Strategy for India and the Developing World. A Twentieth Century Fund Study, Cornell University Press, Ithaca and London, 1976.
G. Myrdal, *Asian Drama – An Inquiry into the Poverty of Nations*, a Twentieth Century Fund Study, Vol. III, Penguin Books, 1968.
Organization for Economic Cooperation and Development, *Development Cooperation*, 1976 Review, Paris, Nov. 1976.
M. D. Richards and W. A. Nielander, *Readings in Management*, South-West Publishing Co., Cincinnati, Ohio, 1958.
L. Squire and H. G. van der Tak, *Economic Analysis of Projects*, a World Bank Research Publication, Johns Hopkins University Press, Baltimore and London, 1975.
UN Industrial Development Organization, *Small-Scale Industry in Latin America*, United Nations, New York, 1969, ID/27.
A. Watterston, *Development Planning Lessons of Experience*, The Economic Development Institute, International Bank for Reconstruction and Development, Johns Hopkins Press, Baltimore, 1965.
World Bank, Agriculture Sector Working Paper, 1972, *Annual Report 1975*.
World Food Council Secretariat, Document WFC/20, Rome.

Index